TOLERANCE DESIGN OF ELECTRONIC CIRCUITS

ELECTRONIC SYSTEMS ENGINEERING SERIES

Consulting Editor   **E L Dagless**
University of Bristol

OTHER TITLES IN THE SERIES

Advanced Microprocessor Architectures   *L Ciminiera and A Valenzano*

Optical Pattern Recognition Using Holographic Techniques   *N Collings*

Modern Logic Design   *D Green*

Data Communications, Computer Networks and OSI (2nd Edn)   *F Halsall*

Microwave Components and Systems   *K F Sander*

TOLERANCE DESIGN OF ELECTRONIC CIRCUITS

Robert Spence

Imperial College of Science and Technology

Randeep Singh Soin

Design Engineering Group, GenRad

Addison-Wesley Publishing Company

Wokingham, England • Reading, Massachusetts • Menlo Park, California
New York • Don Mills, Ontario • Amsterdam • Bonn
Sydney • Singapore • Tokyo • Madrid • San Juan

Cover designed by Crayon Design, Henley-on-Thames and printed by the Riverside Printing Co. (Reading) Ltd. The illustration is based on the photograph of Figure 1.16, showing two implementations of a 43-coefficient (CCD) transversal filter on silicon.
Typeset by Advanced Filmsetters (Glasgow), Limited.
Printed and bound in Great Britain by the Bath Press, Avon.

First printed 1988.

British Library Cataloguing in Publication Data
Spence, Robert
    Tolerance design of electronic circuits. —
    (Electronic systems engineering series).
    1. Electronic circuit design
    I. Title   II. Soin, Randeep Singh
    III. Series
    621.3815′3   TK7867

    ISBN 0–201–18242–4

Library of Congress Cataloging in Publication Data
Spence, Robert, 1933–
    Tolerance design of electronic circuits.

    (Electronic systems engineering series)
    Includes index.
    Bibliography: p.
    1. Electronic circuit design.   2. Tolerance
(Engineering)   I. Soin, Randeep Singh, 1952–
II. Title.   III. Series.
TK7867.S64   1988   621.3815′3   87–35080
ISBN 0–201–18242–4

To our children

Robert and Merin
and
Preetma Kaur

Preface

Every parameter of a manufactured component or circuit has a tolerance associated with it. Thus, the actual value of a 10% tolerance 10 kΩ resistor may be anywhere between 9 kΩ and 11 kΩ. Similarly, the voltage drop V_{BE} across the base emitter junction of a bipolar transistor in an integrated circuit may show a variation of ± 50 mV. For this reason, the performance of a mass-produced circuit is liable to vary from one sample to the next, simply because the parameters of its constituent components exhibit unpredictable variation. If this variation in circuit performance is such that the customer's specifications on performance are violated by some of the manufactured circuits, then only a fraction – called the yield – of the mass-produced circuits may be acceptable to the customer. Clearly, this fraction may become too low to be acceptable for economic reasons, and there is a need to know how such a circuit may be redesigned to reduce these unwanted effects. In particular, primary interest is often placed on redesign to achieve increased – and preferably the maximum obtainable – manufacturing yield. Such redesign is called **tolerance design**.

The subject of tolerance design was first researched during the early 1970s. As a consequence, a substantial number of algorithms for the automatic redesign of a circuit to reduce the statistical variation in its performance and to increase its manufacturing yield were thoroughly tested by the early 1980s. Although research still continues into this fascinating problem, enough is now known about tolerance design, and enough algorithms have been found by experience to offer valuable design aids, that it is now timely to bring tolerance design methods to the attention of industry. That is the principal purpose of this book. It is also expected that it will form a useful introduction to those students who wish to pursue research into this interesting topic.

The origin of the book is to be found in research into tolerance design that was begun in the Department of Electrical Engineering of Imperial College in the early 1970s: by the early 1980s a substantial research team had been built up. The view was then taken that sufficient was known about the commercially important topic of tolerance design that a short course should be made available to industry. It was given by researchers both from Imperial College and from outside organizations, and the response in terms of numbers attending was most encouraging. Since that time the course has been repeated in a number of locations though, intentionally, the number of presenters has diminished and the subject matter has been more carefully integrated. The nature of the content also changed: from being a collection of introductory talks and research studies it became an essentially tutorial presentation. The course notes underwent the same change, and led to the preparation of the present book. It is important, in fact, to stress that this book is not a research monograph primarily directed to those already researching the field, although it may well be useful to new researchers. Rather, it is directed to circuit designers, CAD specialists and design managers in industry in order to acquaint them with what the available tools can achieve.

In preparing this book the authors have benefited tremendously from their association with other workers concerned with tolerance effects. Collaborators in the first courses included Nicos Maratos (now teaching in Athens), Paul Rankin (of Philips Research Laboratories), Ian Jones (now in the USA), Nick James (a software consultant in London), Ajoke Ilumoka (now lecturing in Nigeria), Eric Wehrhahn (Philips Communications, Nurnberg) and Bob Wong (a recent graduate from Imperial College), and we are grateful for the benefit we thereby received. Work with Laszlo Gefferth from Budapest also influenced the final manuscript. In its final stages the book benefited tremendously from a most thorough and detailed critical analysis by Eric Wehrhahn, and for this we are most grateful to him.

Although it can be satisfying and even fun to write a book of this sort, the task is made immensely more rewarding if its authors are supported in this task by their spouses, in this case Kathy and Pavitra. Neither is an expert in tolerance design, and this makes their support and encouragement even more appreciated.

Robert Spence
Randeep Soin

Figure acknowledgements

The authors and publishers are grateful to the IEE and the IEEE for permission to reproduce the following figures from previously published work:

Figures 1.14, 1.15 and 1.16 are taken from Knauer, K. and Pfleiderer, H. J. (1982) 'Yield enhancement realized for analogue integrated filters by design techniques.' *Proc. IEE, Pt. G*, No. 4, 122–126. © IEE 1982.

Figures 1.17 and 1.18 are taken from Ilumoka, A. I. and Spence R. (1980) 'Statistical approach to reduction of circuit variability.' *Electronics Letters*, **16**(20), 761–762. © IEE 1980.

Figures 3.3 and 3.4 are taken from Nassif, S. R., Strojwas, A. J. and Director, S. W. (1986) 'A method for worst-case analysis of integrated circuits.' *IEEE Trans. Computer-Aided Design*, **CAD-5**(1), 104–113. © IEEE 1986.

Figures 3.7 and 4.14–4.18 are taken from Spoto, J. P., Coston, W. T. and Hernandez, C. P. (1986) 'Statistical integrated circuit design and characterization.' *IEEE Trans. Computer-Aided Design*, **CAD-5**(1), 90–103. © IEEE 1986.

Figures 3.12–3.14 are taken from Leung, K. H. and Spence, R. (1977) 'Idealized statistical models for low cost linear circuit yield analysis.' *IEEE Trans. Ccts. Sys.*, **CAS-24**(2), 62–66. © IEEE 1977.

Figures 3.18, 7.22 and 9.14 are taken from Pinel, J. F. and Roberts, K. H. (1972) 'Tolerance assignment in linear networks using non-linear programming.' *IEEE Trans. Ccts. Theory*, **Ct-19**(5), 475–479. © IEEE 1972.

Figure 4.13 is taken from Soin, R. S. and Rankin, P. J. (1985) 'Efficient tolerance analysis using control variates.' *IEE Proc., Pt. G*, **132**(4), 131–142. © IEE 1985.

Figures 4.21, 4.22, 6.4 and 6.5 are taken from Rankin, P. J. (1982) 'Statistical modelling for integrated circuits.' *Proc. IEE, Pt. G*, **129**, 186–191. © IEE 1982.

Figures 7.5, 7.13, 7.23 and 7.24 are taken from Soin, R. S. and Spence, R. (1980) 'Statistical exploration approach to design centering.' *Proc. IEE, Pt. G*, **127**(6), 260–269. © IEE 1980.

Figures 7.20 and 7.21 are taken from Wehrhahn, E. (1984) 'A cut algorithm for design centering.' *Proc. IEEE Int. Sym. Ccts. Sys.*, Montreal, 1984, 970–973. © IEEE 1984.

Figure 8.10 is taken from Singhal, K. and Pinel, J. F. (1981) 'Statistical design centering and tolerancing using parametric sampling.' *IEEE Trans. Ccts. Sys.*, **CAS-28**(7), 692–702. © IEEE 1981.

Figures 8.11 and 8.13 are taken from Antreich, K. J. and Koblitz, R. K. (1981) 'An interactive procedure for design centering.' *Proc. IEEE Int. Sym. Ccts. Sys.*, 139–142. © IEEE 1981.

Figures 9.16–9.18 are taken from Ilumoka, A. I. and Spence, R. (1982) 'A sensitivity based approach to tolerance assignment.' *Proc. IEE, Pt. G*, **129**(4), 139–149. © IEE 1982.

Figure 10.1 is taken from Agnew, D. (1980) 'Design centering and tolerancing via margin sensitivity minimization.' *Proc. IEE, Pt. G*, **127**(6), 270–277. © IEE 1980.

Figures 10.3 and 10.4 are taken from Soin, R. S. and Rankin, P. J. (1985) 'Efficient tolerance analysis using control variates.' *IEE Proc., Pt. G*, **132**(4), 131–142. © IEE 1985.

Contents

CHAPTER 1

The Problem

OBJECTIVES

The aim of this introductory chapter is to describe the problems whose origin can be traced to the tolerances associated with all manufactured components. If an electronic circuit is designed for mass production then, as a consequence of the tolerances associated with all manufactured components, the performance of the mass-produced circuits will exhibit variation, sometimes to the extent that the specifications laid down by the customer will be violated. In such a case the manufacturing yield is less than 100% and may – for a variety of reasons – be economically unacceptable. Such a situation immediately gives rise to a number of questions, all concerned with how to minimize the unwanted effect of component tolerances. Such minimization is referred to as **tolerance design**. No indication is given in this chapter as to *how* these questions can be answered: rather, the objective is to identify the sort of question that will typically be asked by the circuit designer, and to give the reader some idea of the success that can be achieved with the help of the newly developed computer aids to tolerance design. The chapter concludes by illustrating, in concrete terms, the design improvements that can be achieved by the algorithms that form the subject of this book.

1.1   Circuit design

Electronic circuits are designed in response to the requirements of a customer. The customer may, for example, specify the limits of acceptability to the insertion loss of a filter (Figure 1.1), the voltage gain of an amplifier (Figure 1.2), the HIGH and LOW d.c. levels of a logic gate (Figure 1.3) or the time-domain response of a switch (Figure 1.4). Sometimes the specifications are laid down by an international authority such as the CCITT and are rigidly fixed, while at other times they may be negotiable with the customer ('I can manufacture a cheaper circuit if you can relax this part of the specification by 0.1 dB'). The latter situation is quite common when the customer happens to be a fellow designer in the same company.

In response to the customer's request, the circuit designer will propose a first trial design whose origin may be his own experience, a rough calculation on the back of an envelope, a book of tables or an earlier circuit designed for a similar function and specification. However, rather than construct a physical prototype to test this first design, it is likely that the designer will simulate it on a computer. At this stage it might not be surprising if (say) the predicted insertion loss (Figure 1.5) violated the customer's specification over certain bands of frequency. By some means – which may exploit sensitivity calculations, optimization techniques or the designer's own insight – and possibly in the course of a number of iterations, the circuit will gradually be modified until its performance eventually meets the specifications (Figure 1.6). At this stage the designer may feel that some measure of success has been achieved.

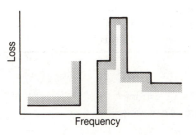

Figure 1.1
Specifications on the performance of a filter.

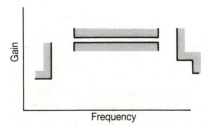

Figure 1.2
Specifications on the performance of an amplifier.

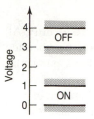

Figure 1.3
Specifications on the d.c.
performance of a logic gate.

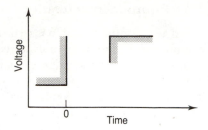

Figure 1.4
Specifications on the time-domain
performance of an inverter.

1.2 A depressing scenario

Let's suppose that the customer wants to purchase 10 000 samples of
the circuit that has been designed to meet the specification. On the basis
of the result shown in Figure 1.6 the circuit may be put into mass
production with some confidence. Nevertheless, if the manufactured
circuits are individually tested at the end of the production line it may
well be found that, within the first 100 samples produced, only 60 meet
the specification. In other words, up to this point, the manufacturing
yield is 60%.

The immediate response to this situation will almost certainly be
to halt production, since the economic viability of the design has been
put in question. The next move might be to measure the responses of all
100 circuits and plot them on the same graph (Figure 1.7). The result
shows a considerable spread of the circuit performance around the
'designed' performance shown in Figure 1.6.

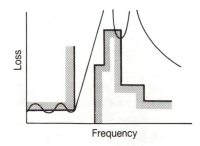

Figure 1.5
A nominal circuit that violates the
specifications.

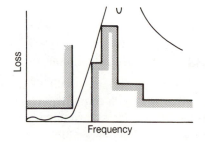

Figure 1.6
A nominal circuit that satisfies the
specifications.

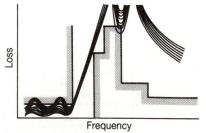

Figure 1.7
Some samples of a mass-produced
circuit violate the specifications.

1.3   Component tolerances are responsible

The reason for the spread in insertion loss is not difficult to identify. All components within the manufactured circuit have tolerances associated with them, so that the value of each component will, for a particular sample circuit, differ to some random extent from its nominal value. Thus, a resistor with a nominal value of $100\,\Omega$ and a tolerance of 10% will actually have a value between 90 and $110\,\Omega$: the next similar resistor taken from the box of '$100\,\Omega$' resistors will also have a value between 90 and $110\,\Omega$, but in all likelihood will be different from the previous one. For some components the parameters characterizing them may be subject to an even greater tolerance: the current gain of a transistor, for example, might only be guaranteed to lie between 20 and 300. Since the component values of each manufactured circuit are in general different from the ones designed, it is to be expected that each circuit's performance will usually differ from that (Figure 1.6) of the simulated nominal circuit.

1.4   Questions

In the scenario just depicted, the manufacturing yield of a circuit was found to be 60%. Severe economic consequences are obviously associated with this situation since, without some remedy, the unsatisfactory circuits (which must not reach the customer) must either be thrown away or, if possible, repaired in some way. The latter course of action could, in fact, turn out to be very costly both in time and in use of skilled manpower, and could easily prove more expensive than merely throwing the unsatisfactory circuits away.

Thus, with a manufacturing yield as low as 60% – indeed, with any yield less than 100% – a number of questions may be voiced by the designer who, while being fully aware of the underlying cause of the unsatisfactory yield, may well be at a loss to know what corrective action can and should be taken.

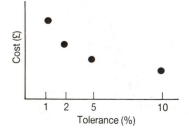

Figure 1.8
Component cost usually decreases as the tolerance increases.

1.4.1   Tolerance adjustment

Since the unsatisfactory yield is caused by component tolerances, and it is known that with zero tolerances (that is, with the nominal circuit) the specifications are not violated (Figure 1.6), a proposal to tighten the component tolerances appears, at first sight, to be reasonable. However, it is usually the case that the cost of a component is a strongly inverse function of its tolerance (Figure 1.8) and therefore, while the yield may be increased, the circuit will cost more to manufacture. The question therefore arises as to whether there is some optimum choice of component tolerances for which the two effects balance out. More precisely,

Question 1

Is there some optimum choice of component tolerances which, though perhaps leading to less than 100% yield, nevertheless minimizes the cost of acceptable circuits?

1.4.2   Rogue components

Let us, for the moment, consider keeping the component tolerances fixed, so that the total component cost of the circuit doesn't change, and let us investigate the effect of changing the *nominal* value of one or more components. It may well be the case, for example, that the major cause of the unsatisfactorily low yield is *one* particular resistor whose nominal value is too high. In other words, those manufactured circuits in which this resistor happens to have a high value (but one still within its tolerance range) tend to exhibit unacceptable performance. For this reason the designer may ask

Question 2

Is the yield particularly sensitive to the nominal values of any components? If so, which components are they?

1.4.3   Adjustment of nominal values

If the answer to the first part of Question 2 is 'yes', then the designer will additionally wish to ask

Question 3

What are the optimum nominal values of the components which will lead to maximum manufacturing yield, and what is the value of that yield?

1.4.4   Negotiable specifications

In some cases it may be difficult or impossible to alter component tolerances or nominals; in others the improvement in yield obtained by adjustment of these parameters may be too small to cause a significant reduction in cost.

In such situations an exercise in improving the manufacturing yield by altering the specifications may prove to be profitable. This may at first sight appear to be an unusual step to contemplate, for it seems like cheating. There are many situations, however, where it will be justified, since there is almost always an arbitrary element in performance specifications. For example, it is not uncommon for specifications to have the form that they do because they have been copied from previously commissioned designs rather than having been derived from a detailed analysis and knowledge of the present application. Thus, for a frequency-selective circuit, the customer may, on reflection, agree to relax the specification at certain sensitive frequencies if the designer identifies these parts of the specification as having an inordinate effect on the yield.

Yet another commonly occurring situation is one in which the particular circuit being investigated performs one function within a system which includes other circuits performing different functions. Whereas the specifications on the overall system function may be fixed, it may be the case that this does not imply a unique set of specifications on the constituent circuits. The system designer (who in this case is the customer) will then try to negotiate and distribute the specifications among the constituent circuits, taking into account achievable yields, related specifications and the need to minimize total cost.

A slightly broader exercise than the one described above will not confine itself to the negotiation of a relaxation of specifications to which the yield is particularly sensitive, but also offer the customer, in return, a tightening of certain other specifications which are easier to meet or which have an insignificant effect on yield. Thus, the designer may ask

Question 4

How sensitive is the yield to the various performance specifications? What is the relative difficulty of meeting different specifications? What changes in specification should I attempt to negotiate with the customer?

1.4.5   Product improvement

The specifications on circuit performance may be set by the designer's own company. The aim may be to manufacture and offer for sale not just a power supply, but a power supply that is better in some way than those offered by its competitors. One measure of quality might be the limits within which the performance can be guaranteed to the purchaser. A product with more stringent limits on the possible spread of performance will be preferred.

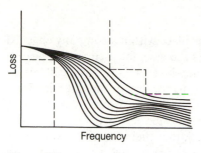

Figure 1.9
Original performance spread.

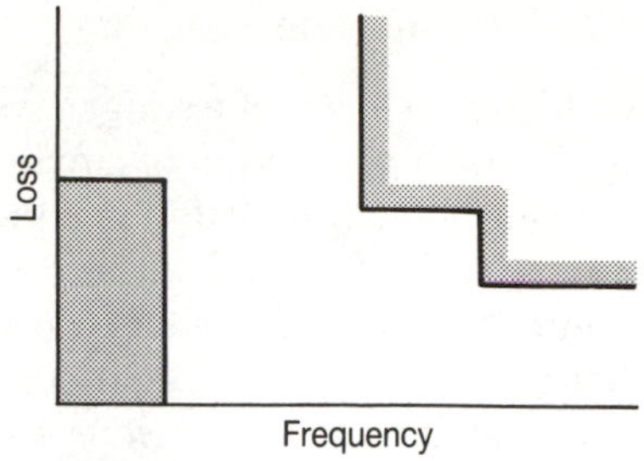

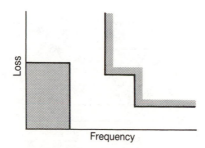

Figure 1.10
Published tolerances.

For a particular design the yield is determined by the spread in performance over a large number of manufactured circuits and the specification limits imposed by quality control. For an early design the spread in performance (Figure 1.9) and the yield achieved may be such that the performance limits that can be quoted (Figure 1.10) render the circuit marketable. Nevertheless, it may be possible to redesign the circuit by a choice of different nominal component values such that a smaller performance spread ensues (Figure 1.11), allowing tighter performance limits (Figure 1.12) to be imposed without compromising yield. In this way a more attractive product may be offered to potential purchasers. These considerations may be summarized as

Question 5

Is it possible to redesign my circuit economically so that the performance variability over a satisfactory fraction of the manufactured samples is reduced?

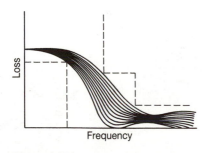

Figure 1.11
Improved (reduced) performance spread.

Figure 1.12
Improved published tolerances.

1.4.6   Component screening

At least two forms of testing may take place within a company engaged in the mass production of circuits. As the assembled circuits appear at the end of the production line they may be tested (automatically or by human beings) to see if their performance meets the specifications. Quite separately, components (e.g. resistors, capacitors and transistors) bought in from an external supplier may be tested before their connection within a circuit, either to check that their parameter values are indeed within the stated tolerances or to permit selection according to some criterion laid down by the designer ("we must only use transistors with current gains greater than 50 for this circuit"). This criterion may have been formulated following an answer to Question 3. Clearly, this latter scenario will have relevance only to the manufacture of circuit boards using discrete components rather than the manufacture of integrated circuits.

But testing can be costly and its use, as a consequence, must be carefully justified. One situation in which component testing may be justified is illustrated in Figure 1.13. It may, for example, have been established that, of all the parameters describing the components within a circuit, it is the current gain (β) of a particular transistor to which the manufacturing yield is especially sensitive (Question 2). In this case it may be economically justifiable to measure the current gain of each such transistor, and only connect it within a manufactured circuit if the gain lies within specified limits. Thus, the designer may ask

Question 6

Of all the components in my circuit, which, if any, would it be economic to measure before possible inclusion in the manufactured circuit, with a view to maximizing the manufacturing yield?

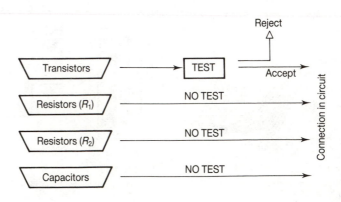

Figure 1.13
Testing of one component that may seriously affect the manufacturing yield.

1.4.7 Other questions

Many other questions may be voiced by the electronic circuit designer in the context of the adverse effect of component tolerances, but we shall not consider them here for good reason. First, because the aim of this introductory chapter is simply to give a flavour of the problems associated with component tolerances. Second, because other questions can more easily be introduced after further discussion.

1.5 Answers

The problems highlighted above cannot, except in the most simple and trivial of examples, be solved directly by a straightforward calculation. Instead, they usually require the attention of complex algorithms and substantial computational facilities. Such algorithms do exist, and form the subject of this book. However, to provide motivation for the reader, we provide two examples which illustrate the application of such tolerance design algorithms.

1.5.1 Yield maximization of an integrated filter

In 1982, Knauer and Pfleiderer of Siemens AG, Munich, West Germany, designed a 43-coefficient transversal filter to be fabricated on a single silicon chip in MOS technology (Knauer and Pfleiderer, 1982). Its circuit schematic is shown in Figure 1.14, its performance specification in Figure 1.15 and its final realization in Figure 1.16, which shows two circuits, for a reason that will soon be apparent. Briefly, the circuit provides, in a reliable, economic and miniaturized form, a filtering function that might otherwise call for a conventional discrete component circuit of considerable complexity and cost.

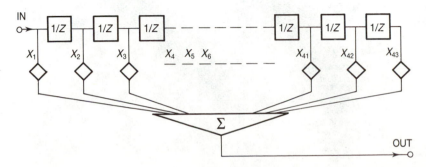

Figure 1.14
A 43-coefficient transversal filter.

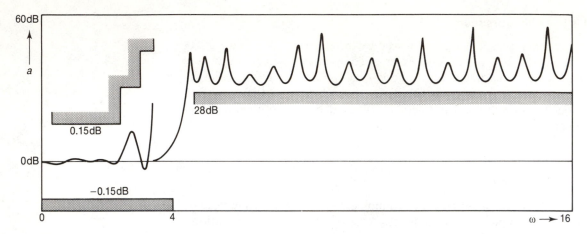

Figure 1.15
Specifications on the response of the filter of Figure 1.14, and the nominal response.

A natural requirement on the design was that any tolerances on parameter values arising from ever-present variations in the integrated circuit fabrication process should have as little effect as possible on the manufacturing yield of the circuit. It should first be emphasized that a first and vital stage in the achievement of this goal involved, not software, but the ingenuity of the designers. The design they eventually created without any computer-based tolerance design aids was fabricated (Figure 1.16), and 600 samples manufactured. Measurements showed that the yield was 65%. The designers then took this design and

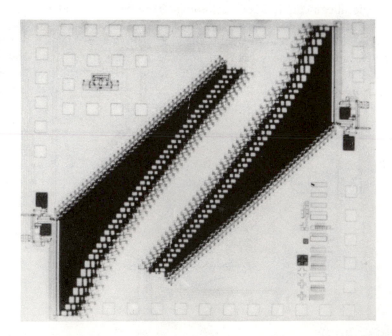

Figure 1.16
Implementation of two versions of the filter of Figure 1.14 on silicon.

subjected it to an algorithm (the centres of gravity algorithm described in Chapter 7): essentially they posed Question 3. The algorithm provided a new set of 43 coefficients defining the components in the circuit, and estimated that the yield would thereby increase to 74%. The improved design was fabricated (that's the second circuit in Figure 1.16), and measurements on the 600 samples showed that this increased yield had been achieved. The two circuits were fabricated on the same chip to offer a reasonable comparison between designs.

1.5.2   A Sallen-Key filter example

Much discussion has been directed to the problem of reducing the performance variability of mass-produced circuits, and was summarized in Question 5. Such discussion has often been directed to a bandpass filter realized by the Sallen-Key circuit configuration (Figure 1.17). This circuit serves as a realistic example to test and compare the effectiveness of different algorithms, since much is known about its sensitivity to parameter changes.

   In 1980, Ilumoka and Spence decided to attack the problem illustrated in Figure 1.18. That is, given the Sallen-Key filter of Figure 1.17 with specified tolerances on the circuit performance, and with the requirement that the Q-factor of the filter should lie within the range from 10 to 40, select nominal values for the component parameters in

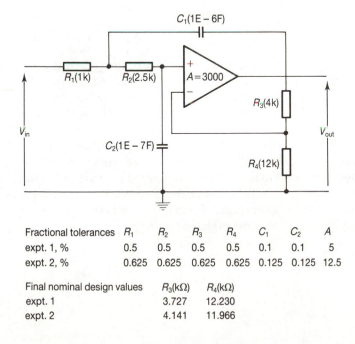

Fractional tolerances	R_1	R_2	R_3	R_4	C_1	C_2	A
expt. 1, %	0.5	0.5	0.5	0.5	0.1	0.1	5
expt. 2, %	0.625	0.625	0.625	0.625	0.125	0.125	12.5

Final nominal design values	$R_3(k\Omega)$	$R_4(k\Omega)$
expt. 1	3.727	12.230
expt. 2	4.141	11.966

Figure 1.17
A Sallen-Key filter.

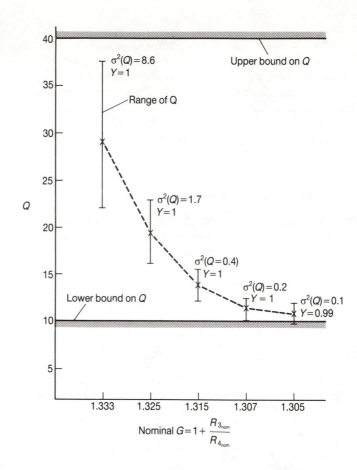

Figure 1.18
Reduction of the variability of the Q-factor of the circuit of Figure 1.17.

such a way as to minimize the variability of the Q-factor. By means of a 'variability reduction' algorithm (Ilumoka and Spence, 1980), the variance of the Q-factor was reduced from an initial value of 8.6 (though with all samples within the specified range for Q; i.e. with 100% yield) to a value of 0.1, the only penalty being a small reduction in yield to 99%.

In the introductory chapters the examples given are those of filter circuits, but there is nothing inherent in the problems addressed and methods discussed to make them more suited to filters rather than other types of circuit. It will be found more convenient at first to discuss discrete component circuits, though the methods and concepts are generally applicable to integrated circuits.

CHAPTER 2

Concepts and Representations

OBJECTIVES

The aim of this chapter is to familiarize the reader with the concepts, representations and terminology needed for an understanding of tolerance analysis and design by discussing and representing in graphical form the problems posed by component tolerances and the means by which they can be solved. Much of the discussion is focused upon **component space**: this is a space having as many dimensions as there are components in the circuit and in which, as a consequence, a single point represents a single circuit. Tolerances on component values thereby define, not a point, but a region in component space called the **tolerance region**, R_T; every sample of a mass-produced circuit will be represented by a point within R_T. While specifications are placed on *circuit performance*, it is the representation of these specifications in component space as the **region of acceptability**, R_A, that is most fruitful when discussing tolerance effects. Tolerance design is, broadly speaking, the action of fitting R_T (defining the manufactured circuits) entirely or partially within R_A (corresponding to acceptable circuits).

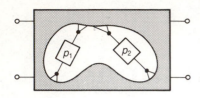

Figure 2.1
A circuit containing two
components.

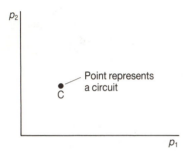

Figure 2.2
A point in parameter space defines a
unique circuit.

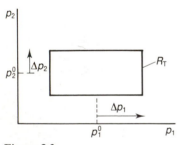

Figure 2.3
The tolerance region (R_T) in
parameter space.

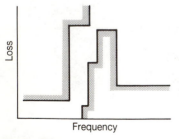

Figure 2.4
The customer's specifications
expressed in performance space.

2.1   Parameter space

To simplify our discussion we shall use as an illustrative example the simple case of a circuit (Figure 2.1) containing only two components. The example is not, however, trivially simple, since the concepts it allows us to discuss generalize, without restriction, to a circuit of any size. For the moment it will be assumed that each component is characterized by a single parameter: for example, a resistor by a resistance. Later, we consider the case where a single component may be characterized by more than one parameter: for example, a transistor by its current gain and cut-off frequency.

If the values of the parameters p_1 and p_2 describing the components of the circuit of Figure 2.1 are known, then the corresponding point C in **parameter space** with axes p_1 and p_2 (Figure 2.2) represents the circuit. *The idea that a single point in parameter space represents a circuit is an important one, and vital to our discussion.*

In practice, we may not know the exact values of p_1 and p_2, since our resistor and capacitor may have been picked at random out of boxes containing many samples having the same **nominal** value. Thus, we may know only that p_1 has a nominal value of p_1^0 and a **tolerance** of plus and minus Δp_1. If the same is true for parameter p_2, then all we can say for certain is that, in parameter space, the point C describing the circuit lies somewhere (Figure 2.3) within a rectangular **tolerance region** (R_T) defined by the nominal values and tolerances of the parameters p_1 and p_2. Indeed, that is *all* we can say without carrying out measurements on the two components. Since measurements cost money, and may not always be possible (with integrated circuits, for example), we shall normally have to operate with the degree of uncertainty inherent in the circuit description of Figure 2.3. To be precise, while a point in parameter space represents a specific circuit, the tolerance region represents the bounds on all possible samples of a mass-produced circuit.

2.2   Circuit performance

As discussed in Chapter 1, the customer provides the specifications on circuit performance, usually by means of upper and/or lower bounds of acceptability on circuit performance (Figure 2.4). Unfortunately, the **performance space** (sometimes called 'output space') in which the specifications are represented is not the same as the parameter space (sometimes called the 'input space') within which we have, readily available, a simple description (R_T) of the bounds within which the manufactured circuits lie. If the specifications were, somehow, to be transformed into component parameter space, the result would be to define (Figure 2.5) a region known as the **region of acceptability** (R_A).

The relation between specifications as conventionally expressed, and their representation by a region of acceptability in parameter space is simply illustrated by the example of a resistive potential divider (Figure 2.6(a)), an example which is sufficiently well understood and simple to be additionally useful in the initial testing of a new tolerance design algorithm. The upper and lower limits on output voltage

$$5.5\,\text{V} > V_{\text{out}} > 4.5\,\text{V}$$

transform respectively into boundaries A and B in parameter space (Figure 2.6(b)). Similarly, the upper and lower limits on input resistance

$$120\,\Omega > R_{\text{in}} > 80\,\Omega$$

transform, respectively, into the boundaries C and D. Together, the four boundaries define, in parameter space, the region of acceptability R_A corresponding to the specifications on the two circuit properties (output voltage and input resistance).

If the point describing a circuit lies within R_A, then the circuit's performance satisfies all the specifications: it is a 'pass' circuit. If it does not, its performance will have violated one or more of the limits which, together, make up the specification: it is a 'fail' circuit. Thus, the circuit described by point W satisfies the specification on voltage gain but fails the specification on input resistance, and is therefore characterized as a fail circuit. The circuit described by point X fails both sets of specifications.

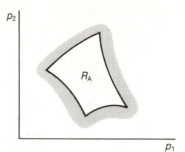

Figure 2.5
The region of acceptability in parameter space defined by the specifications expressed in performance space.

Figure 2.6
A simple circuit, and the region of acceptability R_A in parameter space corresponding to upper and lower specifications on output voltage and input resistance.

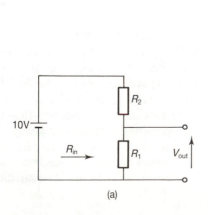

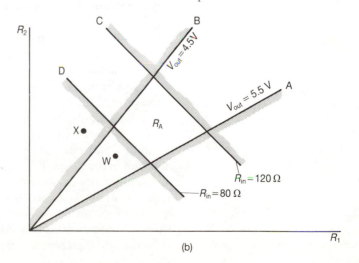

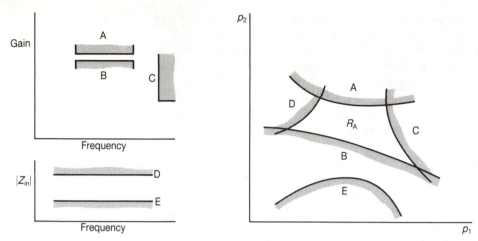

Figure 2.7
Illustrating possible relations between bounds in performance and parameter spaces.

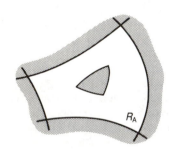

Figure 2.8
A 'black hole' in parameter space.

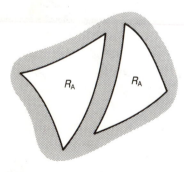

Figure 2.9
Disconnected regions of acceptability in parameter space.

Although it is not the case in this simple illustrative example, R_A is generally quite irregular in shape, with both concave and convex segments. Each segment of the boundary of R_A is normally associated with a unique segment of the specification, as illustrated in Figure 2.7. This figure also illustrates the possibility that a particular segment of the specification (denoted E in the figure) plays no part in the definition of R_A. In other words, whenever a circuit violates that part of the specification denoted by E, it also violates at least one other boundary associated with the specification.

The region of acceptability of a practical circuit can exhibit far more complex shapes even than that illustrated in Figure 2.7. It is possible, for example (Styblinski, 1979), for R_A to surround smaller regions associated with fail circuits (Figure 2.8), or to exist in two or more separate parts (Figure 2.9). In fact, little is known of the shape of R_A for typical circuits and their specifications. Most useful circuits contain at least five, and usually *many* more, components, so that parameter space has many dimensions. It therefore does not lend itself to easy visualization or economic exploration. Several computational procedures may be envisaged which compute the details of R_A from a knowledge of the circuit and its specifications, and thereby derive some interesting two-dimensional cross-sections of R_A. In practice this would involve a search in multidimensional space incurring a prohibitive computational effort. Indeed, it is the huge computational burden involved in the calculation of the boundary of R_A that renders the task of tolerance design so difficult. One possibility that must always be kept in mind is that the combination of a given circuit and specifications on its performance are incompatible, so that whatever values the components take on, the specifications can never be satisfied: in such a case R_A is nonexistent.

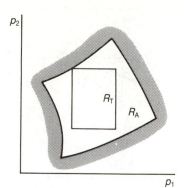

Figure 2.10
The manufacturing yield is 100%
since no circuits fail the
specifications.

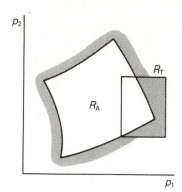

Figure 2.11
Some manufactured circuits may fail
the specifications, so the yield may
be less than 100%.

2.3   Manufacturing yield

If we display both the tolerance region R_T and the region of acceptability R_A in parameter space, useful insight can be gained into the unwanted effects of component tolerances. For example R_T may lie entirely within R_A (Figure 2.10). In this case, since all manufactured samples of the circuit are represented by points which lie within R_T, they must all also lie within R_A. Thus, all manufactured circuits will satisfy the specifications, and the **manufacturing yield** (the fraction of the manufactured circuits that satisfies the specifications) is 1.0 or, as it is usually expressed, 100%.

Sometimes R_T does not lie wholly within R_A (Figure 2.11), in which case any circuit sample represented by a point within the shaded region would fail the specifications, all other circuits being acceptable. In this case the yield would be less than 100%.

2.4   Yield estimation

In practice we will be able only to estimate the yield. The reason is illustrated in Figure 2.12, which shows the result of two separate production runs, each producing 50 circuits which are nominally the same: in other words, each component has the same nominal value and tolerance. Each manufactured sample of the circuit will, as we have seen, be represented by a point in parameter space. In case A, 10 of those points representing manufactured samples happened to lie within the region (external to R_A) associated with 'fail' circuits, so the yield (actual, not estimated) on that occasion was 80% (10 failed and 40

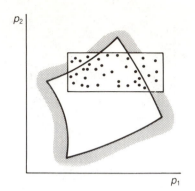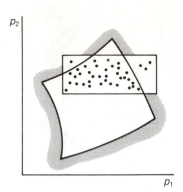

Figure 2.12
Showing how the actual yield can vary between production runs: A (left) and B (right).

passed the specifications). In case B only four of the samples failed, so the actual yield was 92%. The problem in practice is that we do not know beforehand the exact locations of the points describing the manufactured circuits. The component parameters are, in fact, **random variables**. The best we can do is to simulate the random occurrence of parameter values and see what the corresponding yield is: but this value is then only an estimate of the actual yield.

2.5   Parameter distributions

If a large number (say 1000) of nominally identical '10 kΩ' resistors were to be taken from a batch supplied by the manufacturer and individually measured, a histogram showing the number of resistors having actual values lying within selected class intervals might appear as in Figure 2.13(a): the particular histogram shown has the limits corresponding to a 10% tolerance resistor. However, since discontinuous functions do not lend themselves to simple mathematical treatment, it is usually more convenient to discuss a situation in which an *infinite* number of resistors are measured, and where the class intervals are infinitessimally small, so that a continuous **probability density function (pdf)** can be drawn (Figure 2.13(b)).

Interpretation of the continuous curve is straightforward. First, consider the integral of the pdf (Figure 2.13(c)), a curve called the **cumulative distribution function** or **cdf**. With reference to the cdf, the probability that the value of a resistor selected at random will lie in the range R_1 to R_2 is $f_1 - f_2$, where f_1 and f_2 are the values of the cumulative distribution function for R equal to R_1 and R_2. A probability density function is often denoted by the symbol ϕ. To emphasize the fact that it is a function of a parameter (R in Figure 2.13(b)), the dependent variable may be made explicit, as in $\phi(R)$.

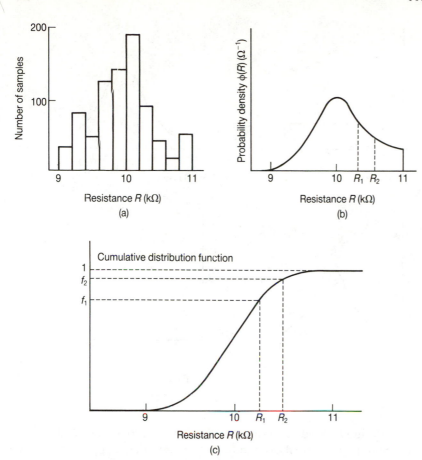

Figure 2.13
Parameter distributions associated with finite and infinite numbers of samples.

Clearly, because *all* the resistor values must, in the example shown, lie between 9 and 11 kΩ, the total area under the pdf must be unity: i.e. there is a probability of one (i.e. certainty) that all the values lie between the limits of 9 and 11 kΩ. Therefore, for a pdf $\phi(x)$ of a random variable, we may say that

$$\int_{-\infty}^{\infty} \phi(x)\,dx = 1 \tag{2.1}$$

Since the performance of a circuit depends on the value of each component parameter, the manufacturing yield is usually dependent on the pdfs of some or all the components within the circuit. It is for this reason that, in this book, we shall make frequent reference to component pdfs.

The shape of ϕ will obviously depend on the nature of the process by which the components being measured are manufactured,

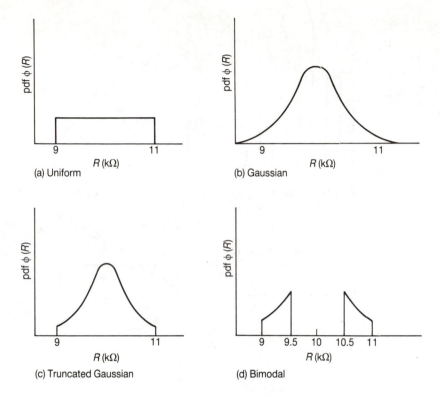

Figure 2.14
Some component probability
density functions.

but certain idealized functions are often referred to, for convenience of
description and analysis and/or because they approximate to
frequently encountered distributions. They include (Figure 2.14) the
uniform distribution where all values are equally likely, the *Gaussian
distribution* which theoretically extends without limit, the more realistic
truncated Gaussian distribution, and a *bimodal* type of distribution
which can so easily result if, for example, all resistors within a middle
range are preselected for sale as (more expensive) 5% resistors.

The Gaussian (often called the **normal**) probability density
distribution function, shown in Figure 2.15(a), is described by

$$\phi(x) = \frac{1}{\sigma\sqrt{(2\pi)}} \exp\left[-\frac{(x - \mu)^2}{2\sigma} \right]$$ **(2.2)**

where x is the random variable, μ is its **mean** and σ^2 its **variance**. For
any random variable the mean or **average** is given by

$$\mu = \int_{-\infty}^{\infty} x\,\phi(x)\,dx$$ **(2.3)**

and its variance by

$$\text{Variance} = \sigma^2 = \int_{-\infty}^{\infty} (x - \mu)^2 \, \phi(x) \, dx \qquad \textbf{(2.4)}$$

The variance is a measure of the dispersion of the random variable about its mean. We are often interested in the positive square root (σ) of the variance, a quantity called the **standard deviation**. If the pdf is

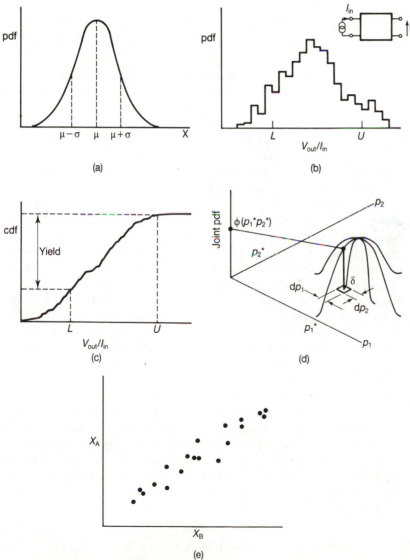

(a)

(b)

(c)

(d)

(e)

Figure 2.15
(a) The Gaussian distribution.
(b) The use of a pdf to characterize the distribution of samples of transfer impedance $V_{\text{out}}/I_{\text{in}}$.
(c) The cumulative distribution related to (b), and used to read off values of yield. (d) Illustration of a joint pdf relating to two parameters R_A and R_B. (e) Scatter plot of pairs of resistances over a number of chips.

associated with a component parameter, the standard deviation is a measure of its tolerance. For example, for a component with a Gaussian distribution, it is a reasonable engineering approximation to take the tolerance to be $\pm 3\sigma$. In practice only a finite number (n) of components are available for measurement, in which case the mean and variance of a sample of resistors, each characterized by a resistance R_i, may be estimated using the following formulae:

$$\hat{\mu}_R = \frac{1}{n} \sum_{i=1}^{n} R_i \tag{2.5}$$

$$\sigma_R^2 = \frac{1}{n-1} \sum_{i=1}^{n} (R_i - \hat{\mu}_R)^2 \tag{2.6}$$

Exactly the same approach to the characterization of spread – by means of a probability density distribution partly or wholly described by a mean and a variance – is applied to circuit performance. Our example of a resistor can be thought of as a measure of the voltage response to current excitation at the same terminal pair: with a circuit we may be dealing with the output voltage due to a current applied at the input, but the characterization of the way in which their ratio varies over a collection of circuits can proceed in the same way as for the simple resistor (Figure 2.15(b)). We shall choose circuit performance to illustrate one valuable use of the cumulative distribution function. Normally there are acceptable limits to (in other words, specifications on) the value of a particular circuit performance (for example, L and U in Figure 2.15(b)), and we wish to know the probability of finding a sample of circuit performance lying between the limits L and U: this probability would, of course, be the yield if no other limits existed. The required probability is the area under the pdf of Figure 2.15(b) between the limits L and U, but to save us the problem of measuring an area we simply read off the relevant values of the cdf (Figure 2.15(c)) at L and U and subtract them to find the required area/probability.

Since most circuits of interest have more than one component that is subject to random variation between samples, we need the concept of a **joint probability density distribution**, or **joint pdf**. For illustration we again choose a two-parameter circuit which is easy to visualize but which generalizes to any number of parameters (Figure 2.15(d)). For any given elemental region δ of size $\mathrm{d}p_1$ by $\mathrm{d}p_2$ in the $p_1 p_2$ plane, the probability of finding a combination of p_1 and p_2 in this particular region is given by the area (volume) under the pdf curve bounded by δ. In other words,

$$\text{Probability}\,(p_1, p_2 \in \delta) = \int_{\delta} \phi(p_1, p_2)\,\mathrm{d}p_1\,\mathrm{d}p_2 \tag{2.7}$$

In this case the probability density, plotted 'vertically' in Figure 2.15(d), would have the dimension of per ohm^2.

   An additional effect arises when a number of component parameters are not only subject to random variation, but exhibit **correlation** between their variations. For example it is known that, as a result of inevitable variations in the integrated circuit fabrication process, resistors deposited on a chip will vary considerably in value. However, although two resistors fabricated in *close proximity* will therefore exhibit variation from one chip to another, their ratio will tend to remain within very tight bounds. In other words, samples of their values taken from a random selection of chips might appear as in Figure 2.15(e). A measure of the relationship between two distributions (e.g. of X_A and X_B in Figure 2.15(e)) is the **covariance**, as explained below.

   A measure of how two random variables (X_A and X_B of Figure 2.15(e), say) vary together is the covariance $\sigma_{A,B}$ defined by the integral

$$\sigma_{A,B} = \int_{-\infty}^{\infty} \int_{-\infty}^{\infty} (X_A - \mu_A)(X_B - \mu_B)\phi(X_A, X_B)dX_A dX_B \quad \textbf{(2.8)}$$

the interpretation of which is perhaps enhanced by comparison with Equation 2.4

$$\sigma^2 = \int_{-\infty}^{\infty} (x - \mu)^2\,\phi(x)\,dx \qquad\qquad\qquad \textbf{(2.4)}$$

which defines the variance of a *single* random variable. Another convenient measure is the correlation coefficient denoted by ρ and defined for the case of two random variables X_A and X_B as

$$\rho_{A,B} = \frac{\sigma_{A,B}}{\sigma_A \sigma_B} \qquad\qquad\qquad\qquad \textbf{(2.9)}$$

From a finite sample of pairs of values of X_A and X_B, the correlation coefficient $\rho_{A,B}$ may be estimated using the expression

$$\hat{\rho}_{A,B} = \left[\frac{1}{n}\sum_{i=1}^{n}(X_{A_i} - \mu_A)(X_{B_i} - \mu_B)\right]\Big/\hat{\sigma}_A\hat{\sigma}_B \qquad \textbf{(2.10)}$$

From this expression it can be appreciated that if X_A tends to be high (or low) when X_B is high (or low), then the summation will be greater than if this tendency were absent. The correlation coefficient always lies within the range from -1 to $+1$

$$-1 \leqslant \rho \leqslant 1 \qquad\qquad\qquad\qquad \textbf{(2.11)}$$

Complete correlation between X_A and X_B will lead to a value of $\rho_{A,B}$ equal to ± 1 and independence to a value of zero.

For purposes of illustration we have considered only two component parameters, whereas the concept of a joint pdf characterized by a covariance and a correlation coefficient generalizes without restriction to realistic circuits containing many components. The extension of the concept of a joint pdf to more than two random variables is straightforward. The pdf itself is denoted by $\phi(p_1, p_2, p_3, \ldots, p_K)$, and is a K-dimensional function satisfying the equality

$$\int_{-\infty}^{\infty} \cdots \int_{-\infty}^{\infty} \phi(p_1, p_2, p_3, \ldots, p_K) dp_1 dp_2 dp_3 \ldots dp_K = 1 \qquad \textbf{(2.12)}$$

which is a generalization of Equation 2.1. In an integrated circuit many of the component parameters p_i will be correlated. These dependencies between component parameters taken two at a time are expressed by elements of a $K \times K$ **variance–covariance matrix** $[S]$. In this matrix each *off-diagonal* element S_{ij} is the covariance σ_{ij} between the ith and jth components as defined in Equation 2.8. Each *diagonal* element S_{ii} is the variance of the ith component as defined in Equation 2.4. As an example, the K-dimensional Gaussian pdf will be given by

$$\phi(p_1, p_2, \ldots, p_K) = \frac{1}{\sqrt{|S|} \sqrt{(2\pi)}} \exp \left[\frac{-(\mathbf{P} - \boldsymbol{\mu})^{\mathrm{T}} (\mathbf{P} - \boldsymbol{\mu})}{2|S|} \right] \qquad \textbf{(2.13)}$$

where $\mathbf{P}$ is the vector of component parameters ($\mathbf{P} = p_1, p_2, p_3, \ldots, p_K$), $\boldsymbol{\mu}$ is the vector of average component values ($\boldsymbol{\mu} = \mu_1, \mu_2, \mu_3, \ldots, \mu_K$) and $|S|$ is the determinant of the variance–covariance matrix.

In two situations above we have used the Gaussian distribution as an illustrative example, and there are two good reasons for this. First, it is completely and economically characterized by its mean and

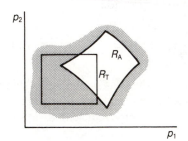

Figure 2.16
Original circuit design.

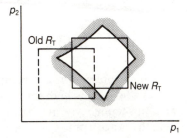

Figure 2.17
Design centring to maximum yield.

variance (and, for more than one variable, the covariance) and lends itself to mathematical operations such as differentiation. The second reason is that the Gaussian distribution tends to occur naturally in connection with manufactured electronic components. However, the methods of tolerance analysis and design to be discussed in this book are not restricted (unless, in rare cases, so stated) to specific component and performance probability density functions.

2.6   Design improvement

If R_T does not lie wholly within R_A (as in Figure 2.11) and the manufacturing yield is therefore less than 100%, what can we do to the circuit – other than starting from scratch and redesigning it completely – to reduce the unwanted effect of component tolerances?

A simple means of increasing the manufacturing yield is suggested by Figures 2.16 and 2.17. It is to adjust the *nominal* values of the parameters, while leaving their tolerances fixed, so that R_T is more centrally located within R_A (Figure 2.17). Such an adjustment of the parameter values to increase the manufacturing yield is called **design centring**.

Having centred the circuit, and achieved a greater yield, it is appropriate to direct attention to parameter tolerances. If these are open to choice, the designer will be aware that a tightening of tolerances (Figure 2.18) will lead to 100% manufacturing yield, though at a higher cost to the customer since the cost of a component is typically an inverse function (Figure 2.19) of its tolerance. Indeed, for this very reason, it may be more economical (Figure 2.20) to perform a trade-off by widening the parameter tolerances so that the penalty of decreased yield is more than offset by the reduced cost of the components. In fact, it is not difficult to appreciate that there may exist an optimum set of parameter tolerances which ensures that the unit

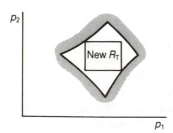

Figure 2.18
A tightening of tolerances leads to increased yield.

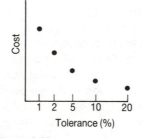

Figure 2.19
A typical relation between component cost and tolerance.

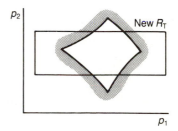

Figure 2.20
Tolerance assignment may lead to lower cost.

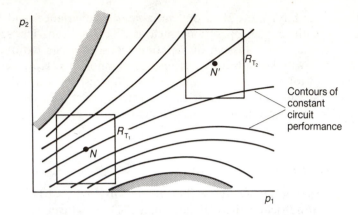

Figure 2.21
Variability reduction.

cost of each satisfactory manufactured circuit is minimized. The choice of parameter tolerances with a view to decreasing the cost of pass circuits is called **tolerance assignment**.

For convenience, the activities of design centring (variable nominal, fixed tolerance) and tolerance assignment (fixed nominal, variable tolerance) have been separately introduced and discussed, whereas they could be regarded as extremes of a combined process in which both nominals and tolerances are simultaneously adjusted.

The last facet of tolerance design we shall introduce in this chapter is called **variability reduction**. The tolerance region R_{T_1} in Figure 2.21 represents an initial design of a circuit containing two toleranced parameters. The contours shown in Figure 2.21 represent loci of constant circuit performance. If the nominal design is now moved from its original location N to a new location N' (Figure 2.21), while keeping the parameter tolerances fixed, the new tolerance region is that shown as R_{T_2}. It is seen from the figure that circuit performance variability will be less with the design represented by R_{T_2} than with R_{T_1}. If the variability in performance from one manufactured circuit to another is reduced, a product operating to tighter specifications can be offered to a purchaser. With reference to Figure 2.21, the adjustment of the nominal design from N to N' to reduce performance variability is termed variability reduction.

Design centring, tolerance assignment and variability reduction are three facets of tolerance design, each requiring a computer algorithm for its successful execution. Researchers have directed much of their attention to design centring, but each facet is considered in one or more of Chapters 7, 8 and 9 which describe various approaches to tolerance design. First, though, since some means of estimating manufacturing yield is essential to any approach to tolerance design, Chapter 3 discusses the principal approaches to tolerance analysis.

CHAPTER 3

Tolerance Analysis

OBJECTIVES

At that stage of circuit design at which the effect of component tolerances must be taken into account it is normal, first, to carry out a **tolerance analysis**. This chapter provides an overview of representative methods of tolerance analysis, the objective of which is to predict the effect on circuit performance of the tolerances inevitably associated with manufactured components. The outcome shows the extent to which the circuit performance will vary from one manufactured circuit to the next, and will provide an estimate of that fraction of the manufactured circuits which satisfies the specifications provided by the customer. This fraction, expressed as a percentage, is known as the **manufacturing yield**. There is no single means of performing tolerance analysis; what this chapter does is to classify and summarize the major available methods and describe, in outline, their advantages and disadvantages. One particular method of tolerance analysis which can also form the basis of effective methods of tolerance design is identified; it is known as Monte Carlo analysis and is discussed in detail in Chapter 4.

3.1   Tolerance analysis and tolerance design

There are two complementary aspects to the consideration of component tolerances and their effect on the performance of circuits. The first, **tolerance analysis**, helps to answer the question: what effect will the component tolerances have on circuit performance? The other, **tolerance design**, tries to answer the complementary question: what can we do to *reduce* the unwanted effect of component tolerances? There are two reasons for considering tolerance analysis first. One is that, unless the extent of the effect is known, it is difficult to know which of a number of facets of tolerance design might be appropriate. The other is that every method of tolerance design has to be based, implicitly or explicitly, on a corresponding method of tolerance analysis.

Chapters 3 and 4 are devoted to tolerance analysis. In this chapter we take a broad view of the available methods, and provide both a classification and a summary of their salient features. From this overview we come to the conclusion that the **Monte Carlo** or **statistical exploration** approach to tolerance analysis is the most robust and useful of the available techniques. Most of the tolerance design methods described in this book are, in fact, based on information generated by a Monte Carlo analysis, and Chapter 4 is devoted to this technique.

3.2   Tolerance analysis objectives

Starting with some knowledge of the statistical variation of component values due to random variations in manufacturing processes, tolerance analysis seeks to predict certain aspects of the resulting variation in

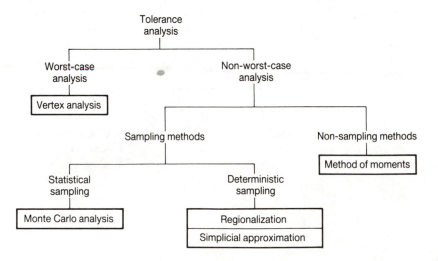

Figure 3.1
Classification of tolerance analysis methods. Boxes identify examples discussed in the text.

circuit performance. For example, the most general tolerance analysis task would be to predict, on the basis of the probability density functions of all components within a circuit, the corresponding probability density functions of all performances of interest. In practice a more limited objective may be set; for example, to estimate the manufacturing yield.

A number of approaches to tolerance analysis are available, and the principal ones are included within the broad classification presented in Figure 3.1. At the first level are to be found **worst-case** and **non-worst-case** methods. Worst-case methods will be treated first.

3.3   Worst-case tolerance analysis

In the absence of any more sophisticated tolerance analysis facilities, the method most commonly used by circuit designers employing circuit simulation packages has been based on the so-called worst-case approach. Its basis is the identification of the extreme (i.e. worst) values of performance resulting from the variations in component values. Because only the extreme performance values are of interest, it follows that the detailed nature of the probability density functions of neither the component values nor the circuit performance values are of any account in such an analysis.

The worst-case tolerance analysis procedure involves, first, the identification of those combinations of component values that cause extreme values of circuit performance. Then a conventional circuit analysis must be carried out for each of the identified component combinations. In general there will be a number of worst-case component combinations, each one associated with one or more performance specifications.

Let us suppose there are m upper specifications and n lower specifications:

$$f_j(P) \geqslant L_j \qquad j = 1 \ldots m \tag{3.1}$$

$$f_k(P) \leqslant U_k \qquad k = 1 \ldots n \tag{3.2}$$

where the f_j and f_k are performance functions, L_j and U_k are the corresponding lower and upper limits of performance acceptability, and P denotes a general set of component values. To perform a worst-case analysis we need to find up to m points $P_1, P_2 \ldots P_m$ in component space such that

$$f_j(P_j) > f_j(P) \qquad \text{for all } P$$
$$j = 1 \ldots m \tag{3.3}$$

and up to n points such that

$$f_k(P_k) < f_k(P) \qquad \text{for all } P$$
$$k = 1 \ldots n$$

(3.4)

We have said 'up to' m or n points for good reason; not all the P_j and P_k will be different in inequalities 3.3 and 3.4 since the same point (i.e. the same combination of component values) may lead to the extreme values of more than one performance function.

3.3.1   Vertex analysis

The main difficulty associated with the worst-case approach is that of *identifying* the worst-case component values; there is no sure method that will apply to all or even most performance functions. An obvious suggestion is to explore the extreme values of the components within their tolerance ranges, as illustrated in Figure 3.2 for the case of three components. The combinations constitute, in fact, the vertices of the tolerance region, and are described by

$$P = p_i^0 \pm t_i$$

(3.5)

where p_i^0, $i = 1$ to K are the coordinates of the nominal point. Unfortunately, there are two main drawbacks to such an approach. One which requires little discussion is the fact that the extreme value of circuit performance may occur, not at a vertex of the tolerance region, but at some location within the tolerance region such as the point A in Figure 3.2.

The other main drawback associated with the assumption that extreme performance values occur at vertices is the computational effort involved in analysing the circuit at the vertices to obtain the corresponding extreme performance values. There are two extreme parameter values associated with each component, so that a circuit of K components is associated with a tolerance region having 2^K vertices. For an unusually small circuit containing 10 components there will be 1024 vertices; for typical circuits there will be millions of them. Clearly, the task of examining each vertex by means of a circuit analysis is economically infeasible.

An alternative to the unacceptable approach of examining every vertex is to predict at which vertices the extreme performance values will be encountered. A common method of so doing is based on the calculation of the sensitivities of the circuit performances with respect to changes in component values, normally at the nominal point in component space. For example, suppose we wish to find the vertex

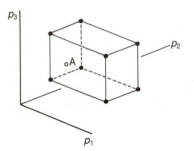

Figure 3.2
Points indicate all combinations of extreme component values. The extreme (worst) value of circuit performance may occur at a point such as A within the tolerance region but not at one of its vertices.

corresponding to the maximum value of a performance function f. First, the sensitivities of f with respect to all component values p are computed:

$$S_i = S_p^f = \left. \frac{\partial f}{\partial p_i} \right|_{P^0} \qquad i = 1 \ldots K \tag{3.6}$$

As before, K is the number of components. Since the effect on a performance function of a small component change depends upon the nominal value of that component, the inclusion of P^0 in Equation 3.6 is justified. The signs of all the K sensitivities are then examined and the K-element vector **X** is formed according to

$$\mathbf{X}_i = \begin{cases} 1 & \text{if } S_i(p^0) > 0 \\ -1 & \text{if } S_i(p^0) < 0 \end{cases}$$

Then, the worst-case for an upper performance specification (e.g. Equation 3.2) is taken to occur at the vertex whose coordinates are given by

$$p_1^0 + X_1 t_1, p_2^0 + X_2 t_2, \ldots, p_K^0 + X_K t_K \tag{3.7a}$$

Similarly the worst-case for a lower performance specification (e.g. Equation 3.1) is taken to occur at the vertex given by

$$p_1^0 - X_1 t_1, p_2^0 - X_2 t_2, \ldots, p_K^0 - X_K t_K \tag{3.7b}$$

In other words, to arrive at the vertex associated with (say) extreme positive performance, one should examine those extreme component values each of which, taken by itself and on the basis of a 'small-change' sensitivity analysis, would lead to a positive change in performance. This procedure would have to be repeated for each performance specification. The justification for this procedure is further explored in Section 3.5.

For the worst cases of performance to be located at the vertices so identified, the relation between circuit performance and component values must obey certain conditions such as monotonicity. Unfortunately, there is no straightforward procedure for testing whether monotonicity is obeyed. Also, as we remarked earlier, extreme performance may not be associated with vertices of the tolerance region. There must always be some uncertainty, therefore, in a worst-case analysis carried out as described above, even with various enhancements that researchers have been able to make (see, for example, Bandler *et al.*, 1975). Nevertheless, useful results *can* be obtained from a worst-case analysis carried out in the manner described above. Also, it need not be expensive computationally; as will be shown in Chapter 9, calculation of the sensitivities of circuit performance to changes in all component values can be inexpensive for d.c. and a.c. performance.

How useful is the concept of worst-case analysis to tolerance *design*? In its identification of the vertices of the tolerance region likely to be associated with extreme performance values, it is of little use in cases where 100% yield either cannot be obtained, or (see Question 1, Chapter 1) where yield is being traded off against cheaper (wider tolerance) components. Nevertheless, the *sensitivity information* generated for use within worst-case analysis may well be relevant to the redesign of a circuit in order to increase the yield, since it indicates the direction in which the nominal design can be moved to reduce the deviation from nominal of the extreme performance values. Agnew (1980) has, in fact, proceeded along these lines with a method of tolerance design exploiting the concept of 'margin sensitivity'.

3.3.2   Worst-case analysis applied to integrated circuits

More recently Nasif *et al.* (1986) have systematically tackled the problems of worst-case analysis applied to the design of integrated circuits. They suggest that worst-case design is most useful in the intermediate stages of design where the expenses of full tolerance analysis carried out for each design stage would be prohibitive. In other situations where the integrated circuit being designed is very large it may not be possible to carry out tolerance analysis for the entire circuit. However, since many large digital integrated circuits are made up from just a few building blocks or cells used many times, it may be possible to carry out worst-case analysis of generic cell types and extrapolate to that of a larger proportion of the integrated circuit. In other cases where, for example, the timing is of major concern, critical paths including the relevant parasitic elements may be identified and worst-case analysis carried out for these paths.

The traditional approach to the worst-case analysis of integrated circuits is as described above for discrete component circuits, but now the component parameters $p_1, p_2 \ldots$ etc. are MOSFET device parameters such as threshold voltages and transconductances. One major weakness of this approach when applied to integrated circuits is that, by working with these electrical parameters, no account is taken of the fact that they are correlated. For this reason traditional worst-case analysis gives over-pessimistic results. An improved approach which applies to certain types of performance function such as signal delays and power dissipation is to identify extreme (worst-case) component parameter combinations from measurements on test devices. Circuit analysis is then carried out with these combinations to obtain the worst-case values of the performance functions. For example, in NMOS technology, four sets of parameters corresponding to 'slow' and 'fast'

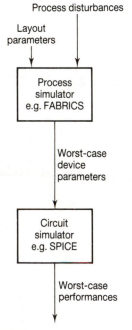

Figure 3.3
Outline of one approach to the worst-case analysis of integrated circuits.

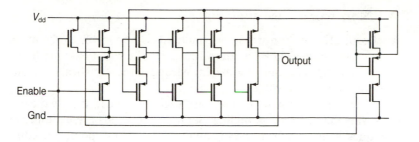

Figure 3.4
A CMOS VCO circuit example for worst-case analysis. See results in Table 3.1.

enhancement and depletion MOSFETs are employed. Nevertheless, this only partially addresses the problem of correlation between component parameters. A more satisfactory approach, propounded by Nasif *et al.* (1986), is based on treating process parameters as the basic component parameters. That is, starting from worst-case *process* parameters, worst-case *device* parameters are obtained using a process simulator. The worst-case device parameters are then fed into a standard circuit simulator to obtain the corresponding performance values (see Figure 3.3). For illustration consider the CMOS voltage controlled oscillator (VCO) shown in Figure 3.4. The performance function of interest is the oscillation frequency, which it is desirable to maximize. Therefore the worst-case analysis is trying to identify the smallest frequency that might occur due to component tolerances. Nasif *et al.* (1986) analysed this example using two approaches: the conventional approach using device parameters (from which electrical parameters are easily obtained), and the approach employing process parameters and a process simulator.

In the first case the device parameters considered were the lengths, widths and transconductances of the N and P channel devices, and the tolerances employed were twice the standard deviations, i.e. $\pm 2\sigma$. In the second case the process parameters considered were: linewidth variations in the nitride and polysilicon layers, the diffusivity of boron, the linear oxide growth rate, the substrate concentration and oxide interface charge density. The worst-case values for oscillation frequency are shown in Table 3.1. In this example as expected the conventional method gives more pessimistic results.

Table 3.1   Results of worst-case analysis applied to CMOS VCO.

Performance	Minimum oscillation frequency (MHz)
Nominal performance	46.90
Worst-case with respect to device parameters conventional method	27.60
Worst-case with respect to process parameter deviations, new method	37.27

3.4   Non-worst-case analysis

The classification of non-worst-case methods is applicable to the more general case where yield is less than 100%. Unlike worst-case methods, the present category requires some knowledge of the statistical distribution of the component values.

The non-worst-case methods are usefully divided into two categories: **sampling** and **non-sampling**. Sampling methods are explorative techniques which perform circuit analysis at sample points in component space. The sample points may be chosen in a regular and systematic manner as in the regionalization and simplicial approximation methods, or pseudo-randomly ('statistically') as in Monte Carlo tolerance analysis. In the other category, that of non-sampling methods, the only one worthy of consideration in view of its not uncommon use is the method of moments, which is an old established method predating much recent research and has also been employed (though under different names) in other engineering disciplines (Bjorke, 1977).

3.4.1   Non-sampling methods: the method of moments

The method of moments is based on a family of mathematical expressions known as the **transmission of moments formulae**, and is illustrated symbolically in Figure 3.5 where only two of the components within the circuit of interest are explicitly shown and described by the parameters p_1 and p_2. The distributions of p_1 and p_2 are necessarily assumed to be known. For purposes of illustration two circuit properties are shown to be of interest, one a voltage (v) and the other a current (i). The object of the transmission of moments formulae is to predict the distributions of these properties. For purposes of tolerance analysis and design, the most important moments of both the component and performance distributions are their variances (i.e. second moments about the mean), and it is with these variances that the rest of this section is concerned.

TRANSMISSION OF MOMENTS FORMULAE

The transmission of moments formulae are based on Taylor series representations of the performance functions. Consider initially just one performance function (f) and its Taylor series representation truncated after the term involving the first derivatives:

$$f(P) = f(P^0) + \sum_{i=1}^{K} \frac{\partial f}{\partial p_i} \Delta p_i \qquad\qquad \textbf{(3.8)}$$

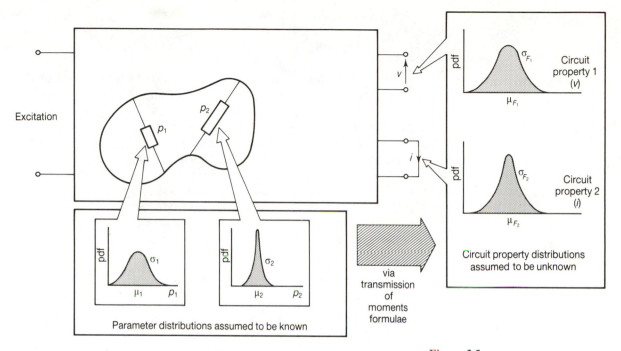

Figure 3.5
Relevant to the discussion of the method of moments.

Here $\Delta p_i = P_i - P_i^0$, $P(= p_1 p_2 \ldots p_K)$ is the point in component space for which the function is being approximated, and $P^0 = p_1^0 p_2^0 \ldots p_K^0$ is the nominal point (the centre of the tolerance region) as before. From Expression 3.8 it is a fairly straightforward task (Hammersly and Handscomb, 1964) to derive the following relationship between $\mathrm{var}(f)$ and $\mathrm{var}(p_i)$, the respective variances of the performance function f and the component parameters p_i:

$$\mathrm{var}(f) = \sum_{i=1}^{K} \frac{\partial f}{\partial p_i} \mathrm{var}(p_i) \tag{3.9}$$

Expression 3.9 is just one of a whole family of transmission of moments formulae which have been used in tolerance analysis and design. This particular formula becomes slightly more complicated if the component parameter values are correlated, as will be the case for integrated circuits:

$$\mathrm{var}(f) = \sum_{i=1}^{K} \frac{\partial f}{\partial p_i} \mathrm{var}(p_i) + 2 \sum_{i=1}^{K} \sum_{j=1}^{K} \frac{\partial f}{\partial p_i} \frac{\partial f}{\partial p_j} \mathrm{cov}(p_i, p_j) \tag{3.10}$$

where $\mathrm{cov}(p_i, p_j)$ is the covariance between component parameters p_i

and p_j. For notational convenience

$$\mathrm{var}(f) = \mathbf{X}[S]\mathbf{X}^{\mathrm{T}}$$

where $\mathbf{X}$, a vector of length K, is the vector of sensitivities of the performance to the component parameters (i.e. $\mathbf{X} = \partial f/\partial p_1, \partial f/\partial p_2 \ldots \partial f/\partial p_k$), $\mathbf{X}^{\mathrm{T}}$ denotes the transpose of $\mathbf{X}$ and $[S]$ is the $K \times K$ variance/covariance matrix. For example element S_{ij} is the covariance between component parameters i and j, i.e. $\mathrm{cov}(p_i, p_j)$; whereas the diagonal elements are simply the corresponding variances $\mathrm{var}(p_i)$, $i = 1 \ldots k$. Thus, using formulae such as 3.9 and 3.10, the variance of circuit performance functions can be computed from knowledge of the variances of the component pdfs. However, we shall not pursue the transmission of moments formulae further, but rather discuss their use in tolerance analysis and design.

TRANSMISSION OF MOMENTS FORMULAE IN TOLERANCE DESIGN

It should be noted initially that the transmission of moments formulae relate the two sets of moments (of component parameters and performance functions) irrespective of the exact form of either component or performance pdfs. We can in fact proceed in two directions, either by *assuming* (for we cannot be sure) a particular form of performance pdf, or by making use of procedures which require no assumptions about the pdfs.

(a) *Gaussian pdfs* The first course, although not strictly necessary, leads directly to more concise results. For a particular form of performance pdf, knowledge of its moments is sufficient to characterize it completely, and the most useful assumption is that the performance pdf is Gaussian. This assumption has the support of the Central Limit Theorem.* Knowledge of the mean and variance of, for example, a single dimension Gaussian pdf completely characterizes it. The calculation of yield also becomes a straightforward task, for it now involves only the integration of a Gaussian performance pdf between two limits, the lower and upper performance specifications:

$$Y = \int_L^U \Omega(f)\mathrm{d}f \qquad\qquad (3.11)$$

where the performance specifications are $L \leqslant f \leqslant U$, the pdf $\Omega(.)$ is

---

*The pdf of a sum of random variables with arbitrary pdfs will tend to be Gaussian, especially if the number of variables is large, the variances of a few variables are not much greater than those of all the others, and the individual pdfs are symmetrical.

given by

$$\Omega = \frac{1}{\sigma_f \sqrt{(2\pi)}} \exp \frac{-(f - \mu)^2}{2\sigma_f^2}$$

and where μ is the mean of f and σ_f^2 is its variance. Integral 3.11 can easily be computed using standard normalized tables (Larson, 1969).

Unfortunately the above discussion has been based on the assumption that only *one* performance function is of interest, whereas in practice there are usually many which simultaneously have to meet specifications for the circuit to be acceptable. The implication for the calculation of yield is that more complicated transmission of moments formulae have to be employed which involve not only the variances of the performances f_i, but also their covariances. Even if these additional moments are calculated, the main problem arises from the fact that the performance pdf $\Omega(.)$ is now multidimensional and the estimation of yield will now involve a multidimensional integration:

$$Y = \int_{L_1}^{U_1} \cdots \int_{L_m}^{U_m} \Omega(f_1 f_2 \ldots f_m) \mathrm{d}f_1 \, \mathrm{d}f_2 \ldots \mathrm{d}f_m \qquad \textbf{(3.12)}$$

where Ω is now an *m*-dimensional Gaussian pdf. Since standard normalized tables do not exist for Gaussian pdfs of higher dimensions, this particular form of pdf is no longer of special value in the above integration. Certain deterministic (rather than statistical) procedures do exist for integrating higher dimension Gaussian pdfs, and the case of yield estimation is thoroughly dealt with by Karafin (1970). Suffice it to say here that the procedures are involved and not easy to implement.

(b) *Arbitrary pdfs* Although the Central Limit Theorem lends general support to the assumption of Gaussian performance pdfs, Pinel and Singhal (1977), in an empirical study of electronic circuits, cast doubts on its use for this problem. We may, therefore, wish to consider procedures which are *independent* of the particular form of the performance pdfs. Such procedures do exist and are based on the so-called Chebychev inequality (Larson, 1969). Put quite simply this inequality, illustrated in Figure 3.6, states that the probability that the value of a random variable will occur less than $\alpha\sigma$ away from its mean is greater than or equal to $(1 - 1/\alpha^2)$ irrespective of the form of the pdf. Here α is an arbitrary constant, σ_f is the standard deviation of f and μ_f is its mean. In other words,

$$\text{Probability} \left(|f - \mu_f| \leqslant \alpha\sigma_f\right) \geqslant (1 - 1/\alpha^2) \qquad \textbf{(3.13)}$$

The Chebychev inequality may easily be used to estimate yield in the following way. Assume, for simplicity, that the lower and upper

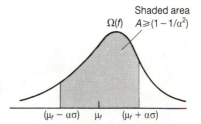

Figure 3.6
A simple illustration of the Chebychev inequality for a one-dimensional pdf.

specifications are symmetrical about the mean value of f; in other words, $\mu_f - L = U - \mu_f$. From a knowledge of σ_f (calculated by means of the transmission of moments formulae) we may compute α such that $\alpha\sigma = \mu_f - L$. Then, an estimate of yield Y (integral 3.11) is simply

$$Y \geqslant (1 - 1/\alpha^2) \tag{3.14}$$

The case where these specifications L and U are not symmetrically disposed about μ can easily be accommodated.

As in the previous case the above discussion has been based on a single performance function. The Chebychev inequality *does* extend to higher dimensions, but with added complexity (Seth and Roe, 1971; Godwin, 1955). The main criticism of yield estimates such as 3.14 is that, although they do not require any assumption about the form of the component pdfs, the answers they provide are too conservative and insufficiently precise.

The last basic drawback to the method of moments that we wish to mention is that the transmission of moments formulae are based on first-order Taylor series representations of the performance functions. Circuit performances will in general be *nonlinear* functions of the component parameters, and the calculation of moments using formulae based on these representations could therefore lead to considerable error. Nevertheless, despite the drawbacks we have discussed, the method of moments can provide a useful first approximate indication of tolerance effects.

THE METHOD OF MOMENTS APPLIED TO INTEGRATED CIRCUIT DESIGN

Recently, Spoto *et al.* (1986) have described a powerful suite of programs for performing statistical analysis, which includes among its facilities both the method of moments and Monte Carlo analysis (to be described below). Their experience is that the method of moments can give useful results for many types of performance functions in integrated circuit design especially where the mismatch in component parameters has a substantial effect on the particular performance whose moments are being estimated. In this context, in writing expression 3.10, they differentiate between mismatch of parameters and process parameters:

$$\operatorname{var}(f) = \mathbf{X}_m^T[S_m]\mathbf{X}_m + \mathbf{X}_p^T[S_p]\mathbf{X}_p \tag{3.10b}$$

where the subscripts 'm' and 'p' refer to matched and process parameters respectively. As an example, sheet resistivity would be considered among the process parameters whereas the mismatch between individual resistors – which depends on geometrical variations arising

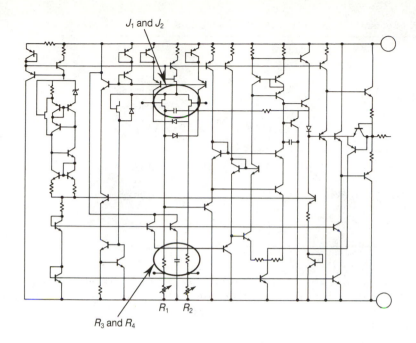

J_1 and J_2

R_3 and R_4

R_1 R_2

Figure 3.7
High-performance, JFET input,
operational amplifier.

from uncertainties in the lithographic processes – are examples of matched parameters. This distinction will become clearer in the circuit example discussed below. First we note that the transmission of moments formula may also be applied to the change of a performance value as a function of a parameter such as temperature. For example

$$\mathrm{var}(\Delta f) = \Delta \mathbf{X}_m^T [S_m] \Delta \mathbf{X}_m + \Delta \mathbf{X}_p^T [S_p] \Delta \mathbf{X}_p \qquad \textbf{(3.10c)}$$

where $\Delta \mathbf{X}_m$ may be the vector of the difference in sensitivities at two different temperatures, and Δf that of the values of their performance function.

The circuit example considered was the high-performance JFET operational amplifier shown in Figure 3.7. The mismatch in the threshold voltage V_p of the two input JFETs and in the resistors R_3 and R_4 largely determine the input offset voltage V_{io}, which was the performance function of interest in this example. The method of moments was used to predict the variance of the offset voltage before trimming (by varying R_1 and R_2) and the variance of the offset voltage temperature variation. In Table 3.2, column two gives the standard deviations (square roots of the variances) of the component parameters. The component parameters, p_1, p_2, p_3 respectively are the percentage mismatch between the threshold voltages V_p of transistors J_1 and J_2, the resistances R_3 and R_4 and the leakage current IDSS. Columns three,

Table 3.2   Results of the method of moments applied to a JFET operational amplifier.

Name	Mismatch $\sqrt{\mathrm{var}(p_i)}$ (%)	Normalized sensitivity $\frac{\partial v_{io}}{\partial p_i} p_i$			Change in sensitivity $\frac{\Delta S_i}{\Delta p_i} = \frac{\partial v_{io}}{\partial p_i}(125\,°C) - \frac{\partial v_{io}}{\partial p_i}(-55\,°C)$
		25 °C	− 55 °C	125 °C	
$p_1 \to \left(\dfrac{V_p(J_1)}{V_p(J_2)}\right)$	0.38	1.33	1.19	1.51	0.32
$p_2 \to R_3/R_4$	1.0	0.138	0.115	0.176	0.061
$p_3 \to \left(\dfrac{I_{DSS}(J_1)}{I_{DSS}(J_2)}\right)$	0.7	0.07	0.06	0.08	0.02

four and five give the normalized sensitivities of the offset voltage to these component parameters at three different temperatures. Column six gives the change in the sensitivities for a change of temperature from − 55 °C to 125 °C.

Using these tabulated values we may now calculate the variance of the offset voltage at 25 °C and the variance of the change in offset voltage over the temperature range − 55 °C to 125 °C. We note first that the correlation coefficient ρ between the mismatch parameters V_p and IDSS is 0.94 while ρ between all the other parameters taken two at a time is negligible. Since covariances can be calculated from ρ using the relationship

$$\mathrm{cov}(p_1 p_2) = \rho_{1,2}\sqrt{[\mathrm{var}(p_1)\,\mathrm{var}(p_2)]}$$

the only covariance that need be considered is that between IDSS and V_p which can be computed as 2.5×10^{-5}. Therefore the variance of V_{io} at 25 °C can be calculated by substituting the relevant values from Table 3.1 into expression 3.10:

$$\begin{aligned}
\mathrm{var}(V_{io}) &= (1.33)^2(0.0038)^2 + (0.138)^2(0.01)^2 + (0.07)^2(0.007)^2 \\
&\quad + 2(1.33)(0.07)(2.5 \times 10^{-5}) \\
&= (5.7\,\mathrm{mV})^2
\end{aligned}$$

Similarly the variance of the change in offset voltage V_{io} over the full temperature range may be calculated as

$$\begin{aligned}
\mathrm{var}(\Delta V_{io}) &= (0.32)^2(0.0038)^2 + (0.061)^2(0.01)^2 + (0.02)^2(0.007)^2 \\
&\quad + 2(0.032)(0.02)(2.5 \times 10^{-5}) \\
&= (1.48\,\mathrm{mV})^2
\end{aligned}$$

As a check on the efficacy of the method of moments Spoto *et al.* (1986) computed the same performance variances using Monte Carlo analysis, since it is generally considered to be the most reliable tolerance analysis method. The corresponding results from a 300-sample Monte Carlo analysis were $(5.6\,\text{mV})^2$ and $(1.5\,\text{mV})^2$. These results suggest that the method of moments may be suitable for evaluating the effect of component mismatch on performance variability.

In terms of computational effort required the method of moments has a great edge over Monte Carlo analysis. For instance in the example above the cost of the former would be in the range of two or three circuit analyses while the Monte Carlo analysis was performed with 300 circuit analyses. However the method of moments, unlike the Monte Carlo method, is not valid for every circuit problem and it is not possible to give *a priori* conditions for its validity for any particular circuit under consideration.

3.4.2 Sampling methods

Sampling methods are based on circuit analysis at sample points in component parameter space. The sample points may be selected in a regular and/or systematic ('deterministic') manner as in the regionalization and simplicial approximation methods, or pseudo-randomly as in the Monte Carlo method. We deal first with the deterministic methods.

DETERMINISTIC EXPLORATION: REGIONALIZATION

Conceptually, an obvious approach is to lay a regular grid over the tolerance region (Figure 3.8), and to perform a circuit analysis at one representative point within each subregion so formed. In the illustration of Figure 3.8 it is the centre of each rectangular subregion that has been chosen to define the circuit to be analysed, and whose performance is then tested against the specifications. All points within a particular subregion are assumed to be pass or fail depending on whether the circuit defined by the central point is correspondingly a pass or fail. The result of carrying out a circuit analysis at each centre point will then be (Figure 3.9) an approximation to that part of the region of acceptability that falls within the tolerance region. Also, in the special case in which the component parameter pdfs are uniform and uncorrelated, an approximation to the yield can be obtained by dividing the number of subregions characterized as pass by the total number of subregions. This approach to tolerance analysis is called **regionalization** (Scott and Walker, 1976; Leung and Spence, 1974).

To see how the regionalization approach can provide an estimate of yield for the more complex case in which the component pdfs are

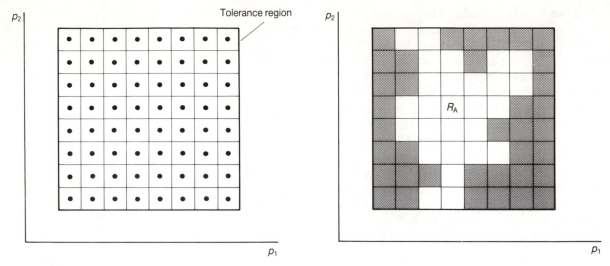

Figure 3.8
Circuit performance at the centre of each subregion is assumed to characterize *all* points within that subregion.

Figure 3.9
Approximation to the region of acceptability obtained from circuit analyses carried out at the points shown in Figure 3.8.

non-uniform and may additionally be correlated, we consider an example involving just two parameters (Figure 3.10). Necessarily, we assume that the joint pdf of the components $\phi(p_1, p_2)$ is known. First, by reference to $\phi(p_1, p_2)$, a very large number of points are generated within the tolerance region by a pseudo-random generator governed by the pdf $\phi(p_1, p_2)$, *though without any corresponding circuit analyses being performed.* The result, obtained at little cost in view of the absence of circuit analyses, is shown in Figure 3.10(a); this could represent, for

Figure 3.10
(a) Generation of 100 points within the tolerance region according to the component pdfs. (b) The number of points lying within each subregion. The weight of a subregion is found by dividing the inscribed number by 100, the total number of generated points.

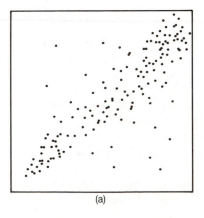

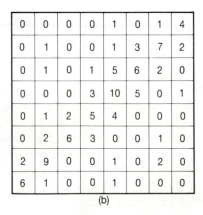

0	0	0	0	1	0	1	4
0	1	0	0	1	3	7	2
0	1	0	1	5	6	2	0
0	0	0	3	10	5	0	1
0	1	2	5	4	0	0	0
0	2	6	3	0	0	1	0
2	9	0	0	1	0	2	0
6	1	0	0	1	0	0	0

(a)

(b)

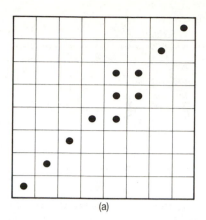

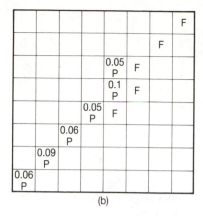

(a)   (b)

Figure 3.11
(a) Points (circuits) selected for analysis on the basis of the corresponding subregion having, in Figure 3.10(b) a weight $\geqslant 0.04$. (b) The results (P = pass, F = fail) of analyses carried out at the points indicated in Figure 3.11(a). For passing circuits, the weight of the region is indicated. Yield is obtained by adding the weights of the pass regions ($= 0.41$).

example, a batch of 100 circuits taken from the production line. Now consider the tolerance region to be divided into subregions, and the number of points lying within each subregion to be counted (Figure 3.10(b)). The numbers obtained, divided by the total number of points generated, are referred to as the 'weights' of the subregions.

There are two ways in which the estimation of yields can proceed. In one (Figure 3.11(a)), those subregions having weights greater than or equal to some threshold value (e.g. 0.04) are selected, and a circuit analysis carried out at their centres. By testing the predicted performances against specifications, each of the selected regions can be classified as pass or fail (Figure 3.11(b)). Yield can now be estimated as the sum, over all the pass subregions, of the weights of those regions. Stated more generally, the yield can be expressed as

$$Y \approx \sum_{\substack{\text{all} \\ \text{selected} \\ \text{subregions}}} w_r \cdot g_r$$

where w_r is the weight of a subregion and g_r, the **test function** associated with a subregion, is 1 or 0 according to whether its representative circuit passed or failed the specifications.

A more precise estimate of the yield can be obtained by two obvious modifications to the above scheme. First, all regions can be taken into account irrespective of their weights. Second, a more precise calculation of the subregion weights can be carried out by the generation of additional points either before or after the analyses.

Some appreciation of the accuracy associated with the regionalization approach can be gained from the example (Figure 3.12) of an 11-element filter in which four elements are described by Gaussian pdfs truncated at $\pm 2\sigma$ points which correspond to $\pm 1\%$ tolerance

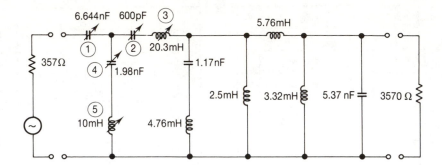

Figure 3.12
Filter circuit subjected to statistical analysis.

limits. For a frequency of 36.16 Hz Figure 3.13(a) shows the pdf of output voltage as obtained from a conventional 5000-sample Monte Carlo analysis, whereas Figure 3.13(b) shows a pdf obtained by the regionalization technique using six class intervals per component (and therefore involving $6^4 = 1296$ analyses). Another comparison is provided in Figure 3.14 where two cumulative distributions are shown. One is obtained from the 5000-sample Monte Carlo analysis, and the other is obtained via regionalization with only four class intervals per component (involving 256 analyses).

The main disadvantage of the regionalization approach is to be found in the manner in which the computational cost increases with the number of component parameters. If each parameter's tolerance range is divided into three regions (and this would be regarded as a very coarse division), and there are K parameters, then the number of subregions is 3 raised to the power K. Even if K is as low as 10, the number of subregions so created would be 59 049; barring exceptional cases, this number of circuit analyses would be regarded as completely uneconomic. The rapid rise in the number of subregions with the number of parameters is often referred to as an instance of the 'curse of

Figure 3.13
Comparison of performance histograms generated by (a) 5000-sample Monte Carlo analysis and (b) the regionalization method, six intervals per component.

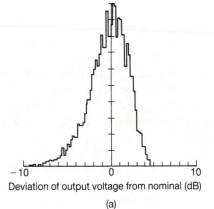

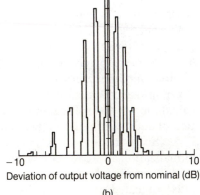

Deviation of output voltage from nominal (dB)

(a)

Deviation of output voltage from nominal (dB)

(b)

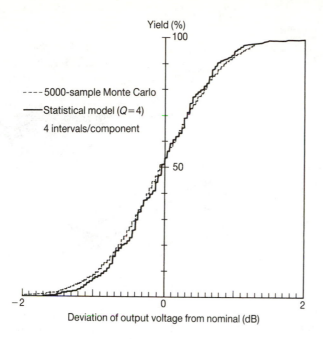

Figure 3.14
Comparison of cumulative distribution curve of output voltage obtained by 5000-sample Monte Carlo analysis with that obtained from the regionalization method.

dimensionality'; that is, a prohibitive increase in the required computational effort with increase in circuit dimensionality. One way to ameliorate this problem is to identify and examine only those regions having a weight greater than some threshold value, as explained above. Another approach, though valid only for linear circuits, is to employ a special algorithm which considerably reduces the cost of each circuit analysis (Leung and Spence, 1975).

Due to the curse of dimensionality mentioned above, the concept of regionalization does not appear attractive for use in tolerance design. The only exception might lie in the domain of integrated circuits where, according to some workers (Hocevar *et al.*, 1983), the number of independent process parameters is very small and can form the basis of tolerance design.

DETERMINISTIC EXPLORATION: SIMPLICIAL APPROXIMATION

An alternative deterministic approach to tolerance analysis, and one that has been more successful than regionalization, is based on the concept of *simplicial approximation* (Director and Hachtel, 1977; Director *et al.*, 1978). Basically, it involves a piecewise-linear approximation to the boundary of the region of acceptability in multi-dimensional parameter space. The computation entails the determination of sufficient points on the boundary of the region of acceptability in

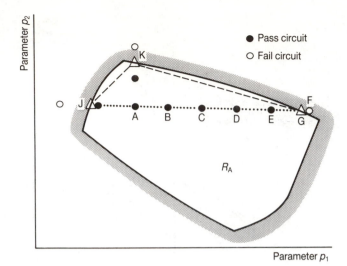

Figure 3.15

A search, involving only circuit analyses, to find the boundary of the region of acceptability. The triangle GJK is a first approximation to R_A.

parameter space to allow a polyhedral approximation to that region to be developed. We first review the technique in outline in two dimensions, then discuss briefly the potential it offers for tolerance design. One characteristic of the method is that, at the outset, no consideration need be given to the location of the tolerance region.

Suppose (Figure 3.15) that a circuit represented by point A in parameter space has been designed, and that simulation by means of a suitable circuit analysis package has shown that it satisfies all performance specifications: it is therefore called a pass circuit. Being a pass circuit, it is known to lie inside R_A. A search for the boundary of R_A is now carried out by varying one parameter (p_1 in Figure 3.15) while maintaining all others fixed; at each step in the search (i.e. at points B, C, D, E and F) a circuit analysis is carried out to determine whether the circuit is a pass or fail circuit. Since the circuit passes at E but fails at F, interpolation of the performances whose boundary has been encountered establishes a point (G) which, to a good approximation, lies on the boundary of R_A. A similar search is now undertaken in the opposite direction to identify the point J as being on the boundary and (say) along the positive p_2 direction to identify point K. The polygon defined by the points G, J and K is taken as a first approximation to the region of acceptability.

As with the example shown in Figure 3.15, a triangular approximation of a two-dimensional region of acceptability is unlikely to be satisfactory. To effect an improvement the following steps may be taken. First, the largest possible circle is inscribed within the polygon (Figure 3.16(a)). Of the sides of the polygon it touches, the longest (JG) is identified, since this is likely to be the poorest approximation to the

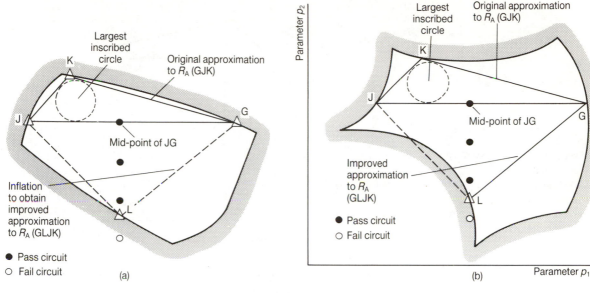

Figure 3.16

(a) Inflation of the first approximation to R_A to include a new point (L) identified by a search from the mid-point of the largest side of the first approximation.

(b) The same construction of an approximation to R_A as shown in (a), but with different specifications, showing how concavity affects the nature of the approximation.

boundary of R_A. From the mid-point of this side, and perpendicular to it in a direction away from the interior of the polygon, a new search direction is defined (Figure 3.16(a)). A search similar to the ones that identified points G, J and K is now carried out to find another point (L) on the boundary of R_A. The original polygonal approximation to R_A is now expanded to include the new point L. This process can be repeated as many times as necessary to obtain the required approximation to R_A: we shall refer to this approximation as R'_A. It will be appreciated that, in Figures 3.15 and 3.16, a two-parameter example has been chosen for ease of illustration. In a circuit with many toleranced parameters the algorithm is essentially unchanged; now, however, R'_A is a multidimensional polygon which, at each iteration, is expanded about the side of greatest area tangential to the largest inscribed hypersphere.

Having obtained such a piecewise-linear approximation to R_A, note is now taken of the location of the tolerance region R_T (Figure 3.17). From knowledge of the probability density function $\phi(p)$ of the parameters p_1 and p_2, sample points are generated within R_T, *but without performing any analysis of the circuits represented by these points*. It is a simple matter to determine, for each point, whether it lies within or without the approximation R'_A to the region of acceptability, and therefore whether it should be classified as a pass or fail circuit. It is then computationally inexpensive to count the number of pass points (N_p) and divide this number by the total number of points generated (N) to obtain an estimate (N_p/N) of the yield. Alternatively, if the

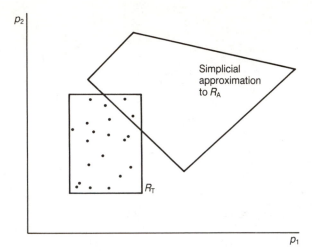

Figure 3.17
Sample points are generated within R_T and tested (at negligible computational cost) to see if they lie inside or outside the approximation to R_A. No circuit analyses are involved.

distributions of p_1 and p_2 are uniform and uncorrelated, the yield can be estimated as the ratio of two volumes:

$$Y = (\text{volume of } R_T \text{ overlapping } R'_A)/\text{volume of } R_T \qquad \textbf{(3.15)}$$

A method for calculating the numerator of Equation 3.15 has been given by Lightner (1979), and is computationally inexpensive.

The major computational effort involved in the simplicial approximation approach to tolerance analysis is associated with the circuit analyses (at points such as A, B and C in Figure 3.15) involved in obtaining the approximation R'_A to the region of acceptability. Later calculations to discover whether sample points lie inside or outside R'_A involve linear programming techniques and are not computationally intensive. In higher dimensions the sides of the polygon become hyperplanes and the polygon itself is referred to as a simplex, hence the name simplicial approximation. Unfortunately the computational cost of a higher dimension simplicial approximation becomes prohibitively high. Indeed, in one circuit example (discussed below) involving only four toleranced components, 307 hyperplanes were required, at a cost of 455 circuit analyses, to give only a moderately accurate approximation to the region of acceptability. The basic reason for this is that the number of hyperplanes required for an *adequate* approximation to R_A rises very rapidly with dimension. Thus, very many circuit analyses have to be performed to find the requisite points on the boundary of R_A. The problem is so severe that the approach is unrealistic for circuits containing more than about five parameters. In other words, just as with regionalization, the curse of dimensionality is encountered.

A major drawback to simplicial approximation is that it requires R_A to be convex and simply connected. Figure 3.16(b) illustrates the problems encountered when R_A is not convex. The simplicial approximation R'_A is not now interior to the true region of acceptability R_A. Unfortunately it is not possible to ascertain whether the R_A associated with a particular circuit example and its performance specifications is convex or not. The requirement of being simply connected means that there can be no 'black holes' in R_A. Unfortunately certain practical circuits (Soin and Spence, 1980; Ogrodski *et al.*, 1980) have been demonstrated to possess an R_A with black holes. Also, compared to Monte Carlo analysis – discussed in the following section – simplicial approximation does not provide any indication of the error involved in its application.

Overall, therefore, we may conclude that the simplicial approximation approach is of limited value in tolerance analysis except when the dimensionality of parameter space is extremely small. Similar remarks apply to the extension of the simplicial approximation approach to tolerance design.

The circuit (Figure 3.18) we choose to illustrate the simplicial approximation method is a fifth-order high pass filter (Pinel and

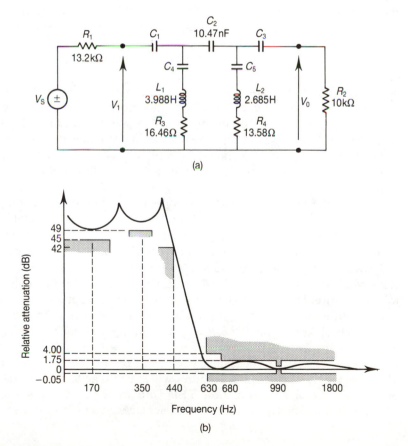

(a)

(b)

Figure 3.18
The circuit (a) and specifications (b) used to illustrate the simplicial approximation method.

Table 3.3 Comparison of yield estimation via the simplicial approximation yield estimation procedure (YEP) and conventional Monte Carlo analysis.

Parameter spread σ (nF) C_1, C_3, C_4, C_5	% yield YEP (455 circuit evaluations)	% yield (500-sample Monte Carlo)
3.5, 3.0, 3.5, 12.5	28	55
1.9, 1.7, 2.2, 6.9	74	90
0.97, 0.85, 1.1, 3.2	100	100

Roberts, 1972) which has served as a test example for many tolerance design methods. The nominal circuit response and the performance specifications are shown in Figure 3.18(b). The results summarized below are more fully discussed by Director *et al.* (1978).

Whereas the original circuit has nine components, the limitation imposed on simplicial approximation by the curse of dimensionality meant that a subset of only four of the nine components was taken to be toleranced. These were the four capacitors C_1, C_3, C_4 and C_5. The simplicial approximation ∂R_A to the true boundary of the region of acceptability R_A, involved 100 vertices and 309 faces and was obtained at a cost of 455 circuit evaluations. Having obtained ∂R_A, yield estimation simply involved the generation of pseudo-random points in component space and the evaluation of that fraction of the total which falls inside ∂R_A, and without the need to perform any additional circuit evaluations. For convenience we refer to this procedure as YEP (yield estimation procedure). To test the method YEP was applied to the high pass filter for three different sets of values of the component variances (hence tolerances) Table 3.3 shows the estimates of yield obtained by this method compared with those obtained from a 500-sample Monte Carlo analysis.

Since the cost of testing whether a point is within or outside ∂R_A is negligible in comparison to that of a single circuit analysis, a suitable comparison of cost would simply involve the total number of circuit evaluations performed. Thus, for the YEP described above, the cost was 455 while that of the comparable Monte Carlo analysis was 500. Returning to Table 3.3, YEP clearly gives poorer results for the higher tolerance, lower yield case. In fact, the accuracy is unacceptably low for most applications. One obvious solution would have been to improve the simplicial approximation ∂R_A by evaluating and incorporating more faces (hyperplanes). However, 309 faces at a cost of 455 circuit evaluations is already rather high. It is for this reason that Director *et al.* developed more efficient procedures which combine features of the original simplicial approximation YEP with those of Monte Carlo analysis. For example, in one enhancement called IYEP (improved

Table 3.4 Comparison of yield estimation via the simplicial approximation improved yield estimation procedure (IYEP) and conventional Monte Carlo analysis.

Parameter spreads (as Table 3.3) C_1, C_3, C_4, C_5	Monte Carlo yield estimations (%) (as in Table 3.3)	% yield IYEP	No. of circuit evaluations	% savings over Monte Carlo analysis
3.5, 3.0, 3.5, 12.5	55	55	$455 + 362 \rightarrow 817$	-63
1.9, 1.7, 2.2, 6.9	90	90	$455 + 131 \rightarrow 586$	-17
0.97, 0.85, 1.1, 3.2	100	100	455	9

Table 3.5 Comparison of yield estimation via the simplicial approximation progressive yield estimation procedure (PYEP) and conventional Monte Carlo analysis.

Parameter spreads (as in Table 3.3) C_1, C_3, C_4, C_5	Monte Carlo yield estimates (%) (as in Table 3.3)	% yield PYEP	No. of circuit evaluations	% savings over Monte Carlo analysis
3.5, 3.0, 3.5, 12.5	55	55	351	30
1.9, 1.7, 2.2, 6.9	90	90	223	55
0.97, 0.85, 1.1, 3.2	100	100	162	68

yield estimation procedure), all the pseudo-random points falling outside ∂R_A are not automatically deemed to be fails. Instead, a circuit evaluation is performed for any such point before deciding whether it is a pass or a fail. This obviously incurs a higher computational cost, as is indicated in Table 3.4. It does, however, achieve the same accuracy as Monte Carlo analysis if the region of acceptability can be guaranteed to be convex.

In a third enhancement called PYEP (progressive yield estimation procedure), yield estimation does not require the prior determination of a simplicial approximation ∂R_A. Instead, ∂R_A is built up progressively as the yield estimation proceeds (Director *et al.*, 1978). The results obtained with PYEP, and presented in Table 3.5, show a saving over Monte Carlo analysis while achieving the same accuracy. However, the reader is reminded that the procedure is limited to four or five components and still assumes convexity of R_A.

STATISTICAL EXPLORATION: THE MONTE CARLO APPROACH

In the Monte Carlo approach to tolerance analysis the sample points in parameter space are generated in a pseudo-random manner to simulate the actual manufacturing process. However, unlike the simplicial ap-

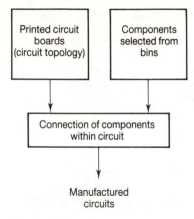

Figure 3.19
The circuit manufacturing process.

Figure 3.20
(a) Generation of N circuit samples
(N points in parameter space)
within the tolerance region.
(b) For each circuit, simulation
determines whether the circuit passes
(●) or fails (○) the customer's
specifications. The yield estimate is
the number of pass circuits divided
by the total number of circuits.
(c) From the results of the
simulations illustrated in (b),
histograms of circuit properties of
interest can easily be derived.

proximation approach, no attempt is made to obtain an approximation to the region of acceptability.

Consider the process of mass-producing a discrete electronic circuit (Figure 3.19). All the manufactured circuits have the same topology (as embodied in the printed circuit board), but the component values vary randomly between the individual circuits. In the case of integrated circuits the component variations will be correlated, and the function of the printed circuit board is replaced by metallization and other forms of interconnection between components. The Monte Carlo method of tolerance analysis directly mimics the process of random component value selection (including the correlations encountered with integrated components) by generating component values according to the known component probability density functions. Figure 3.20(a) illustrates the distribution of sample points generated in parameter space in the course of a Monte Carlo analysis for the case in which p_1 is uniformly distributed over its tolerance range and p_2 is described by a truncated Gaussian distribution function. The N circuit samples generated in this way are then simulated (Figure 3.20(b)) and their performance checked against the customer's specification. Thus, a Monte Carlo analysis is akin to measurements made on N actual manufactured circuits. The manufacturing yield for the N samples can be calculated as the fraction N_p/N of samples that satisfy the specification. If N is sufficiently large (e.g. 200) this yield provides a reasonable estimate of the yield that will be obtained from the actual manufacturing process in which, for example, 10 000 circuits may be produced. In the course of a Monte Carlo analysis histograms of various aspects of circuit performance (Figure 3.20(c)) and various other statistical measures of the performance spread can be generated and are useful to the circuit designer.

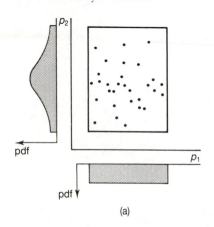

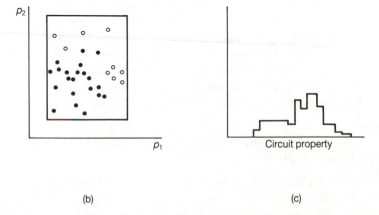

(a)   (b)   (c)

Monte Carlo analysis differs from other methods of tolerance analysis in a number of distinctive ways:

(1)   It is conceptually simple and relatively easily programmed. Given a circuit analysis package, it is a relatively straightforward matter to write a software module that will generate the descriptions of the N circuit samples ready for analysis, and accumulate and process the results of the circuit simulations.

(2)   Although the Monte Carlo estimate of yield is known to be liable to error, the extent of this error can be estimated with known confidence. As we shall show in Chapter 4 it is possible, following a Monte Carlo tolerance analysis, to make a statement of the form

> The estimate of yield is 63%, and we can be 95% confident that the actual yield lies between 51% and 75%.

One can contrast this situation with the simplicial approximation approach for which it is very difficult to obtain an estimate of the error.

(3)   For a given **confidence level** (which was 95% in the quotation above) in the estimate of yield, the **confidence interval** (between 51% and 75% in the example) is proportional to the inverse of the *square root* of the number of samples N. Thus, if 100 Monte Carlo samples (i.e. 100 circuit simulations) led to a confidence interval of $\pm 10\%$ about the estimate, then *four* times the number of samples (i.e. 400) would be necessary to improve the accuracy to the extent of only *halving* the confidence interval to $\pm 5\%$. This property of Monte Carlo analysis is responsible for its substantial but often not prohibitive computational cost. Consequently a number of different modifications to the basic Monte Carlo procedure (Rankin and Soin, 1981; Soin and Rankin, 1985; Hocevar *et al.*, 1983) have been developed to moderate such a dependence of accuracy and cost, while retaining other advantages such as dimensional independence discussed immediately below.

(4)   A mathematical analysis of the Monte Carlo method shows that the relation between the accuracy of yield estimation (the confidence interval) and the number of samples N is independent of the number of parameters subject to random variation. Thus, if the manufacturing yield of a mass-produced circuit is (say) 60%, and one wishes to be 95% confident of the actual yield lying within $\pm 10\%$ of the estimate, then 100 Monte Carlo simulations (i.e. 100 circuit analyses) are required whether the circuit contains

only two or as many as 2000 components. It is this property that allows Monte Carlo tolerance analysis to be employed for medium-size and large circuits for which the computational cost of many of the alternative methods we have discussed would be prohibitively high.

As well as offering an attractive method of tolerance analysis, the result of a Monte Carlo analysis can provide useful clues as to how the circuit design may be modified to increase the manufacturing yield. If we look back at Figure 3.20(b), for example, it appears reasonable to assume that an increased yield might be obtained if the nominal values of both p_1 and p_2 were to be reduced somewhat; then, the fail points would cease to lie in the tolerance region. There is, of course, no guarantee of such a yield increase, since the lower bounds of the tolerance region may lie on or close to the edge of R_A, but at least the possibility of a yield increase may be worth exploring. Indeed, the remainder of this book is based on the conclusion that the Monte Carlo method lends itself well to both tolerance analysis and tolerance design.

3.5   Appendix

The justification behind Expressions 3.7(a) and 3.7(b) as the worst vertices for a particular performance function and specification follows from the following broad argument.

Let us consider the variation in just one performance function $f(.)$ due to the variation in just one component value p_i. Now $p_i^0 \pm t_i$ are the extreme values of p_i that we have to consider, and if the function $f(p_i)$ is assumed to be linear (and hence monotonic) then the value of $f(.)$ at the extreme values of p_i will be given by

$$f(p_i^0 \pm t_i) = f(p_i^0) \pm t_i \frac{\partial f}{\partial p_i}$$

Therefore $f(.)$ will have its largest or smallest value at either of these extreme points, depending on the sign of the sensitivity $\partial f / \partial p_i$.

Let us say we are interested in an upper specification, i.e.

$$f(p_i) < U$$

Then we need to identify the value of p_i that gives the largest value of $f(.)$. Clearly, if the sign of $\partial f / \partial p_i$ is negative, this will be at $p_i^0 - t_i$, and if the sign of $\partial f / \partial p_i$ is positive the largest value of $f(.)$ will occur p_i at

$p_i^0 + t_i$. Defining

$$
X_i = \begin{cases} 1 & \text{if } \dfrac{\partial f}{\partial p_i} > 0 \\ -1 & \text{if } \dfrac{\partial f}{\partial p_i^0} < 0 \end{cases}
$$

we identify the worst-case point for an upper specification to be at

$$
p_i^0 + X_i t_i
$$

A corresponding argument for a lower specification, $f(p_i) > L$, will indicate the worst-case point at

$$
p_i^0 - X_i t_i
$$

If all the components were considered one at a time, we would arrive at the worst-case vertices given by Expressions 3.7(a) and 3.7(b). Now quite clearly this argument makes assumptions about the performance functions which do not hold in most cases. However, in the absence of any other strong criteria for selecting worst-case vertices, this appears to be prudent.

CHAPTER 4

The Monte Carlo Method

OBJECTIVES

Of all the available techniques for tolerance analysis, **Monte Carlo analysis** is not only the most powerful but additionally provides, as a bonus, information that can be exploited in the process of tolerance design. This chapter offers three interpretations of Monte Carlo analysis that are of relevance to both tolerance analysis and tolerance design, and discusses other topics that are vital to the application of Monte Carlo techniques. The discussion of Monte Carlo analysis is sufficiently detailed to permit its understanding and implementation, as well as an appreciation of its limitations. Basically, Monte Carlo analysis is a simulation of the actual manufacturing process, in which the effect of randomly selected component values is assessed computationally, and estimates formed of the fraction of actual manufactured circuits that will meet the specifications. Because it is a statistical method only **estimates** of the effect of component tolerances can be obtained, though it is shown how the degree of **confidence** in these estimates can also be calculated.

4.1   Introduction

In its review of tolerance analysis methods, Chapter 3 presented a description and interpretation of the Monte Carlo method as a technique whereby a random (statistical) exploration of the component space is performed. Whereas Monte Carlo analysis is only one of several different methods of tolerance analysis, we now devote an entire chapter to it for two main reasons. First, almost all the methods of tolerance design discussed in this book are based on the random explorative character of the Monte Carlo method. Indeed, to date, by far the great majority of tolerance design methods devised by various researchers belong to this family. Second, the Monte Carlo method is the most generally applicable, effective and reliable method of tolerance analysis, and is therefore of immense importance for this reason alone.

4.2   Interpretations of the Monte Carlo tolerance analysis method

The name Monte Carlo neither originates from nor is confined to specific methods of electronic circuit design. Far from it; it is used to describe a family of simulation techniques and experimental mathematics of relevance to a very wide range of disciplines. These techniques are characterized by the use and manipulation of pseudo-random numbers within statistical sampling experiments, and are unified by the very general theories and techniques of statistical inference.

When applied in the particular context of the tolerance analysis of electronic circuits, several interpretations of the method are relevant. One view of Monte Carlo analysis is that it is analogous to the pilot run of a circuit-manufacturing process, a process which, in fact, can directly and usefully be simulated by the combination of pseudo-random number generation and a circuit analysis facility. We have already mentioned a second interpretation of Monte Carlo analysis as a statistical exploration of the tolerance region in component space. This view will be seen to be useful when interpreting and developing various methods of tolerance design. Yet a third view of Monte Carlo analysis – as a sampling experiment and therefore subject to the techniques of statistical inference – allows us to compute, for example, the error associated with various Monte Carlo estimates of manufacturing yield. All three interpretations are discussed in this chapter.

4.2.1   Direct simulation of a pilot production run

In tolerance analysis we are concerned with estimating aspects of the statistical variation in performance between manufactured circuits, variations which are induced by statistical variations in nominally identical components. These component variations are, in turn, caused by statistical variations in the processes by which the components are manufactured. Therefore, for the present purpose, the overall manufacturing scheme may be represented as in Figure 4.1.

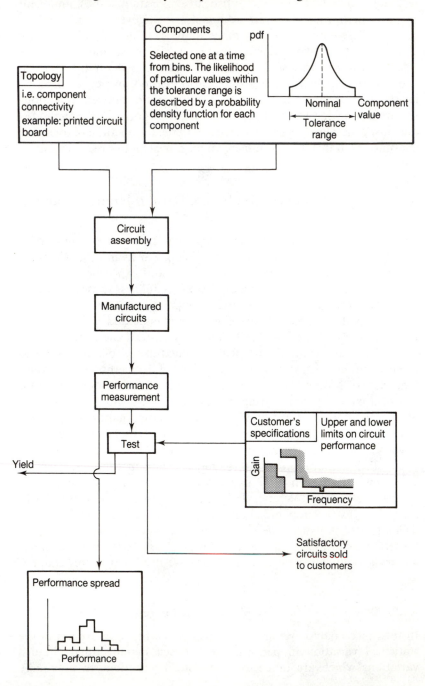

Figure 4.1
The mass production of an electronic circuit.

The act of circuit design consists of a specification, by the circuit designer, of the circuit topology, the component types and nominal values, and (for discrete components) the component tolerances. In the case of discrete components the circuits are assembled on printed circuit boards, the components being individually taken (blindly, without selection) from bins. For integrated circuits the components and their interconnections are fabricated simultaneously in a series of processing steps: statistical variations in these processing steps lead to variations in the parameters of the components and hence in circuit performance. In both technologies, therefore, circuit manufacture may be represented as random component selection followed by circuit assembly though this is only a conceptual representation in the case of integrated circuits.

If the performances of the manufactured circuits are measured (Figure 4.1), the performance values will be found to exhibit deviation from the nominal or 'designed for' values. The extent of this deviation may be such that, when tested, some of the circuits are found to fail to meet the performance requirements ('specifications') provided by the customer. If this is the case, the manufacturing yield will be less than 100%.

With a view to reducing the risk of initiating a production process having a low yield it is of course desirable to obtain an *estimate* of the likely yield before committing a circuit design to full-scale production. One possible approach is to perform a *pilot* production run. Here, a small number of circuits is manufactured, and the yield of the pilot production run calculated as the fraction of manufactured circuits which meet all the performance requirements; this actual yield is then taken as an *estimate* of the yield of the full-scale production run. The number of circuits manufactured in the pilot run must be large enough to give an accurate estimate (otherwise 'freak' samples could give a biased estimate), and yet be much smaller than the total number required from the full-scale run if this method is to be economic. In most cases such pilot production runs would be very expensive and, for integrated circuits, almost as expensive as the full-scale run. Not surprisingly, therefore, this procedure for the estimation of manufacturing yield is not widely used in practice. Conceptually, however, it is: fortunately, simulation not only comes to our aid, but brings with it new benefits.

SIMULATION

The Monte Carlo method directly simulates such a pilot production run according to the scheme shown in Figure 4.2. In this scheme, the circuit is simulated a number of times, but with component parameter values selected via a pseudo-random process which mimics the probability distribution of the physical component parameters.

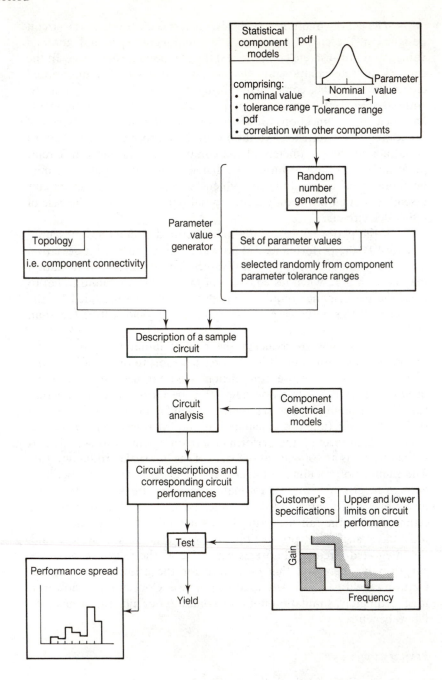

Figure 4.2
Monte Carlo tolerance analysis.

The Monte Carlo tolerance analysis scheme of Figure 4.2 requires a variety of inputs. These include details of the circuit's electrical topology, the component types and corresponding models, and tolerance information in the form of individual probability density functions. In the case of integrated circuits, where the component parameters are correlated, such correlation manifests itself in the form of a joint probability density function. The function of the parameter value generator is to select, via a pseudo-random process, sets of values for the various component parameters subject to tolerance. The pseudo-random process is 'biased' according to the known component pdf, so that each set of values may be treated as if they were measured parameters taken from a randomly selected set of components. The function of the circuit analysis facility is then to predict the performance of a circuit made up from the randomly selected set of component values. The procedure of random component selection followed by circuit simulation is repeated a number (N) of times, and records are kept both of the component value sets and the corresponding predicted circuit performance.

The results generated by the circuit analysis in Figure 4.2 may be treated as if they were the outcome of measurements carried out on physical circuits at the completion of a pilot production run; compare Figures 4.1 and 4.2. Thus, the yield of this pilot run would be found by comparing the predicted performance of each analysed circuit with the customer's specifications, and establishing what fraction of the circuits satisfied the specifications. This yield value would then provide an estimate of the yield of the full-scale production run.

There are two very important differences between the simulated results and those typically obtained from an actual pilot production run. One is that, in addition to measures of performance, the simulation also provides details of the parameter values of the components associated with each manufactured circuit, allowing insight to be gained into the likely effect of parameter adjustments on manufacturing yield (see, for example, Chapters 5 and 7). To obtain this information by measurements on a discrete circuit would be uneconomic, and virtually impossible for an integrated circuit. The other difference is that it is a straightforward matter to study the isolated effect of random variations in (say) the current gain of a transistor; in practice it would be difficult to select a large number of transistors all of whose parameters were identical apart from the current gain.

Monte Carlo tolerance analysis can also be used, not just for yield estimation, but in a more experimental and iterative manner. For example, in the early and often largely exploratory stages of design, the overall function of the circuit may be partitioned by the designer into simpler subfunctions, each to be realized by a separate circuit. Naturally, no specific performance requirements would have been placed on these subfunctions by the customer, so that the question of yield is not directly relevant to the subcircuits. The designer would, nevertheless, have to estimate the performance spreads associated with the various subcircuits, perhaps with a view to initially allocating allowed spreads appropriately among the properties of the subcircuits and then exploring trade-offs. In these circumstances the Monte Carlo method would be most useful for its ability to provide, by pseudo-random sampling, estimates of the various performance distributions.

4.2.2   Monte Carlo analysis as statistical exploration

In a Monte Carlo analysis, each set of K randomly selected component values constituting a circuit may be represented as a single point in K-dimensional component parameter space, while the result of the corresponding circuit simulation is characterized by a point in m-dimensional performance space (Figure 4.3), where m is the number of circuit performance functions of interest. The process of generating pseudo-random points in parameter space and simulating the corresponding circuits to find their images in performance space constitutes 'statistical exploration' of these spaces.

The region of acceptability R_A was introduced in Chapter 2 as the region in parameter space such that, for any point (p) contained in R_A, *all* the performance values $f_1, f_2, \ldots, f_m$ meet the customer's requirements. In other words, the value of the *testing function* $g(p)$ (so called because one is testing a circuit against the customer's specifications) is 1 for $p \in R_A$, and 0 otherwise. Yield was previously defined in terms of $g(p)$ as the integral

$$Y = \int_{R_T} g(p)\,\phi(p)\,\mathrm{d}p \tag{4.1}$$

where ϕ is the joint probability density function of the components. We can provide a simple graphical interpretation (Figure 4.4) of the estimation of yield via the Monte Carlo method for the situation of two toleranced parameters ($K = 2$). First, consider the case where all the

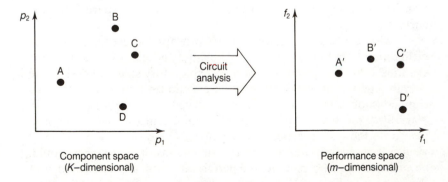

Figure 4.3
Each point in component space defines a circuit, whose predicted performance (defined by a single point in performance space) can be obtained via circuit analyses.

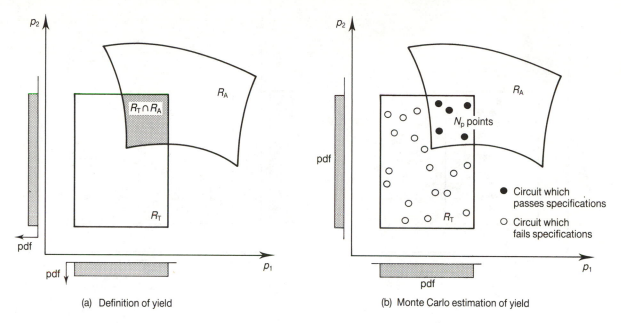

(a) Definition of yield   (b) Monte Carlo estimation of yield

Figure 4.4
Exact calculation of area is replaced by the counting of points to give an estimate of area.

component parameters have uniform distributions as indicated by the pdf plots adjacent to the parameter axes in Figure 4.4(a), and are uncorrelated. In other words, $\phi(p)$ is K-variate uniform over the tolerance region R_T. Yield is then the ratio of the area of the 'overlap' region $R_T \cap R_A$ to the area of the tolerance region R_T (Figure 4.4(a)). In place of this exact calculation, the Monte Carlo method (Figure 4.4(b)) 'sprays' the tolerance region with pseudo-randomly selected points, and for each point the corresponding circuit is simulated and its predicted performance tested against the customer's specifications. If, for a given point (i.e. circuit), all the specifications are met, then the point *must* lie within R_A (whose location we are uncertain of); otherwise, it must lie outside R_A. The manufacturing yield, which is the ratio of the areas $R_T \cap R_A$ and R_T, is then simply estimated as

$$\hat{Y} = N_p/N \qquad\qquad (4.2)$$

where the 'hat' over Y indicates an estimate, N_p is the number of random points falling within R_A and N is the total number of points tested. An important feature of this method of estimating yield is that the boundary of R_A is neither computed nor required.

   For the general case where the component pdf may be other than uniform, the simple graphical interpretation of yield may be extended as shown in Figure 4.5. First, in Figure 4.5(a), we illustrate the generation of the random points for the K-variate situation (here, $K = 2$). For the case where the component pdf is not uniform, two

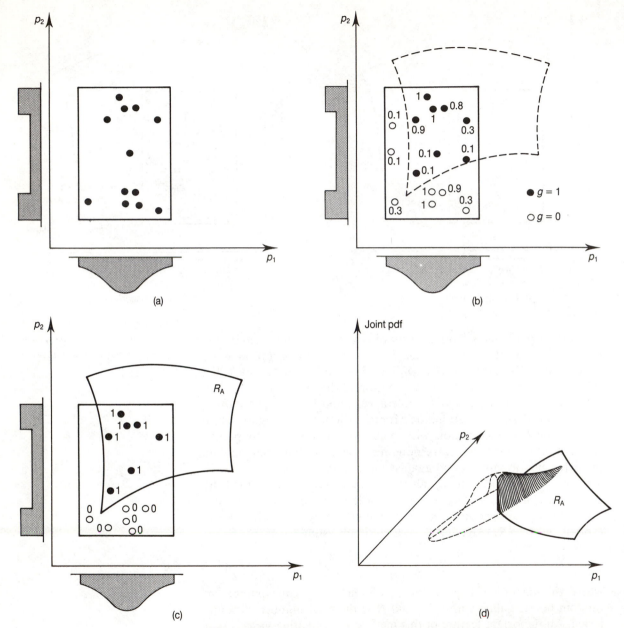

Figure 4.5
Monte Carlo analysis in the situation of a general component pdf. (a) Generation of
random points according to the component pdfs. The pdfs are shown alongside the
relevant parameter axis. (b) Uniform sampling: weights indicate probability of
generation. (c) Sampling according to component pdfs. Testing function is unity for
pass circuits. Directly analogous with actual manufacturing process.
(d) Representation of yield as fractional volume of pdf of R_T coincident with R_A.
The example assumes correlation between the two parameters.

approaches are possible. In one (Figure 4.5(b)) the testing function $g(p)$ for each point is multiplied by a weight (ω_i) which is the probability of a point being generated in that location. The yield estimate is then represented as the ratio of the sums

$$\sum_{i=1}^{N} \omega_i g(p_i) \Big/ \sum_{i=1}^{N} \omega_i$$

Equivalently (Figure 4.5(c)) the points can be generated according to the applicable component pdf, and the yield estimated according to $\hat{Y} = N_p/N$; the latter approach is directly analogous to the actual manufacturing process and the one commonly employed. An interpretation of yield as the ratio of two volumes is provided in Figure 4.5(d); illustrated here is a situation in which component values are correlated.

TOLERANCE DESIGN

One of the principal objectives in tolerance design is to increase the manufacturing yield. Examination of Figure 4.4 shows that, for uniform component distributions, the yield will always be increased if the 'overlap' between R_T and R_A is increased. For component distributions which are *not* uniform, it isn't necessarily the case that increase in the area of overlap will increase yield, although it is usually so. If the customer's performance requirements are not negotiable, so that R_A is fixed, then the yield can be increased (Figure 4.6) by changing

Figure 4.6
The location of pass circuits relative to R_T can indicate possible approaches to the choice of a new R_T with increased yield.
(a) Tolerances remain fixed, and new nominal values lead to higher yield.   (b) A new R_T is formed around the pass points as a first attempt to define a new R_T with higher yield.

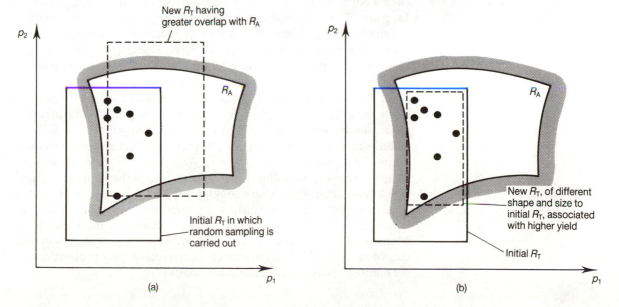

(a)

(b)

the position and/or size of R_T. For example, the yield can be increased by adjustment of the nominal values of the parameters, while maintaining the parameter tolerances fixed (Figure 4.6(a)). In other words, the centre of R_T is relocated. Following such a nominal value adjustment, the nominal may now be fixed and the tolerances then adjusted to achieve higher yield. A combination of these two adjustments is shown in Figure 4.6(b). The extent to which these changes can be made is determined by the relative location and sizes of R_T and R_A, and this information has, to some extent, been explored by the Monte Carlo sampling of parameter space. Thus, the relative locations of pass and fail points established by the Monte Carlo analysis can be used by the methods of tolerance design described elsewhere in the book. Later chapters show in some detail how the spatial information generated by a Monte Carlo analysis can be exploited in tolerance design.

DIMENSIONAL INDEPENDENCE

It is natural to enquire why a systematic (e.g. 'grid') type of sampling within R_T (Figure 4.7(a)) is not preferable to the apparently more complicated statistical sampling (Figure 4.7(b)). The answer lies in a very important facet of statistical exploration, which is its *dimensional independence*. What this means is that, although for both random and systematic sampling the accuracy achieved depends on the number of sample points (N), for a stipulated accuracy the sample size required by random sampling is independent of the dimensionality (the number of circuit components, K). This property is in distinct contrast to what happens if a 'deterministic' exploration is undertaken. With a Manhattan Grid (Figure 4.7(a)) type of search the number of circuit simulations required is Q to the power K, where Q is the number of samples per parameter axis and, as before, K is the number of parameters. Even with very coarse exploration ($Q = 3$, say) and a very small number (10, say) of parameters, the number of points to be evaluated (59 049 in this example) is extremely large and unjustifiable in an economic sense. In a different type of deterministic exploration (called simplicial exploration, see Chapter 3 or 6) an attempt is made to locate the *boundary* of R_A (Figure 4.7(c)), but without restricting the sample circuits to lie within R_T. In this approach the number of circuit simulations required is again roughly exponential in K, to the extent that it is difficult economically to handle more than about five parameters. We see, therefore, that the dimensional independence of Monte Carlo exploration is an *extremely* valuable property, since the cost of an exploration of parameter space is essentially proportional to the number of circuit simulations carried out.

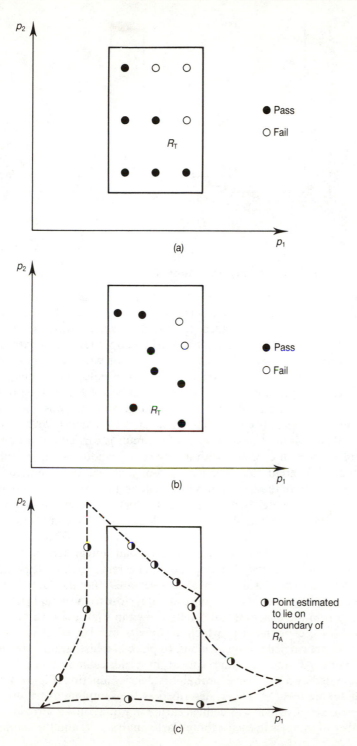

Figure 4.7

Illustration of systematic and random sampling within the tolerance region, and the simplicial approximation approach.

(a) Systematic sampling within R_T.

(b) Statistical sampling within R_T.

(c) Search for points on boundary of R_A. Circuit analyses involved in search are *not* shown. Boundary of R_A shown dashed.

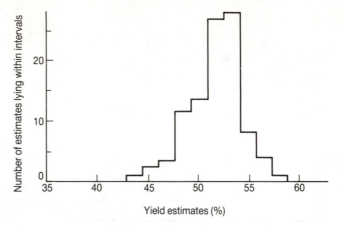

Figure 4.8
The result of one hundred 100-sample Monte Carlo analyses of a given circuit. The (accurate) estimate (51.3%) of the true yield was obtained from a single 10 000-sample Monte Carlo analysis.

4.2.3   Statistical inference and Monte Carlo tolerance analysis

Monte Carlo analysis has been presented, above, as an efficient approach to both tolerance analysis and tolerance design. To be convinced, however, we need to establish such properties as the *accuracy* of the resulting estimate of yield. Properties of this nature are the domain of statistical inference, a subject for which a large body of theory has been developed. Since it is inappropriate to the aims of the book to examine this topic in detail or to prove the results, only a sample of the more significant results will be quoted. Those interested in digging deeper may consult a reference such as Larson (1969).

If the Monte Carlo analysis of a circuit is carried out a number of times, on each occasion with a new set of random samples, then a number of different *point estimates* (y') of yield will be generated. The representation of these estimates by a histogram showing the frequency with which estimates lying in different intervals are encountered will lead to a result such as that shown in Figure 4.8. This histogram was obtained from a real simulation experiment involving 100 Monte Carlo analyses, each of 100 samples, of a bipolar integrated circuit (see Section 4.3) (Soin and Rankin, 1985). The result of this experiment illustrates a typical spread in yield estimates and the fact that the probability is greater of obtaining estimates close to the actual yield.

From the experimental results shown in Figure 4.8 we turn to a *hypothetical* experiment in which an *infinite* number of Monte Carlo analyses are carried out on a circuit. In place of a histogram of the form of Figure 4.8 the resulting spread in yield estimates ($\hat{Y}$) is now represented by a *continuous* probability density function (Figure 4.9) we shall denote by $\theta(\hat{Y})$. This is described as the sampling distribution of the estimator $\hat{Y}$ which was defined earlier in Equation 4.2. Provided the number of samples in each Monte Carlo analysis, N, and the estimated

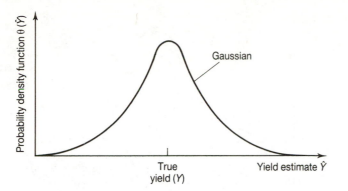

Figure 4.9
The probability density function
$\theta(\hat{Y})$ describing the distribution of
yield estimates obtained from an
infinite number of Monte Carlo
analyses.

yield $\hat{Y}$ obey the approximate condition $N\hat{Y}(1 - \hat{Y}) \geqslant 6$, it can be shown that the curve $\theta(\hat{Y})$ is approximately Gaussian (Van der Waerden, 1969, page 29). For practical values of yield this implies a sample size of about 40. A second useful property of the estimates is that their *average* will tend to the true value of yield Y. This second property, referred to as *unbiasedness* of the estimator, is often expressed as

$$E_{\theta(\hat{Y})} = Y \tag{4.3}$$

where $E_{\theta(\hat{Y})}$ is the expectation or average with respect to the pdf $\theta(\hat{Y})$. These properties are supported by the experiment that led to the results displayed in Figure 4.8.

The two properties of $\theta(\hat{Y})$ just referred to, and which are relevant to an *infinite* number of Monte Carlo analyses, nevertheless allow some useful conclusions to be reached from the outcome (y') of a *single* Monte Carlo analysis. We know that y' is only an estimate, but we shall now discuss how close it might be to the true value of yield, and how this proximity depends on the size (N) of the sample. In fact, following a Monte Carlo analysis leading to a single estimate of yield, we shall be able to make a statement such as

> The yield estimate is 63% and we can be 95% confident that the actual yield lies between 51% and 75%.

In this statement, the values of 51% and 75% are the lower and upper *confidence limits*, and 95% is the *confidence level*. Theories of statistical inference allow us to make such a statement, as we shall now see.

We consider again (Figure 4.10) the theoretical pdf $\theta(\hat{Y})$ whose average is the true but unknown value (Y) of the yield. If two limits symmetrically located about Y are defined as $Y - L$ and $Y + L$ (Figure 4.10), then the probability that any particular point estimate y'

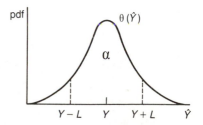

Figure 4.10
The probability of a single estimate
of yield lying between the limits
$Y - L$ and $Y + L$ is the area α
under the curve θ between these
limits.

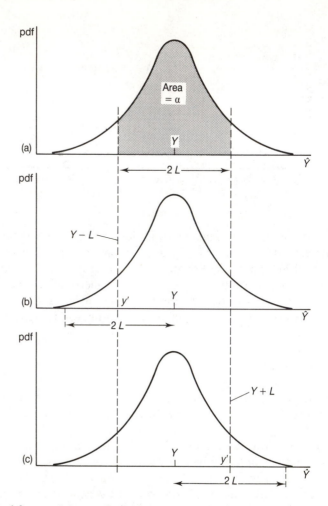

Figure 4.11

If the point estimates y' occur at the limits of the original confidence interval, the new intervals still contain the true value of yield Y. (a) Probability that y' falls within range $Y \pm L$ is α. (b) The condition $y' = Y - L$. (c) The condition $y' = Y + L$.

obtained from a Monte Carlo experiment lies between these two limits is the area α under the pdf between these limits. In practice, however, it is the point estimate y' that is known, and we have no idea of the relative location of the curve $\theta(\hat{Y})$. Let us assume that, without knowing Y, the centre of the curve, we can nevertheless compute $2L$, the size of the confidence interval for any particular value of α. In practice to do this for the Gaussian curve we simply need to know σ^2, its variance. We shall return to this presently.

First let us define an interval, of size $2L$, symmetrically disposed about the observed point estimate y'. In Figure 4.11(b) and 4.11(c) we have taken two extreme cases where the observed estimate y' occurs at the limits of the interval $Y \pm L$. Clearly the confidence interval $y' \pm L$ in these two extreme conditions embraces the true yield Y. Therefore we may conclude that since the probability that y' lies in the interval $Y \pm L$ is α, the probability that the interval $y' \pm L$ contains Y is also α.

Table 4.1   The relationship between L, the size of the confidence interval, and α, the level of confidence.

α	L/σ	α	L/σ
0.803	1.29	0.9	1.65
0.85	1.44	0.931	1.82
0.871	1.52	0.95	2.00
0.89	1.6	0.99	2.6

In other words, commencing with the statement

$$\text{Probability}\,(Y - L < y' < Y + L) = \alpha \tag{4.4}$$

we can easily rearrange the inequality inside the bracket to obtain

$$\text{Probability}\,(y' - L < Y < y' + L) = \alpha \tag{4.5}$$

Therefore, having obtained the point estimate y', we simply need to compute $2L$, the size of the confidence interval, for any particular level of confidence α that we may choose. This then allows us to make statements about our confidence in the yield estimate.

For a Gaussian pdf the relationship between L and α depends on its variance σ^2. In practice tabulated values for a normalized Gaussian pdf with a variance of unity are available and a sample set of values is shown in Table 4.1.* Thus all that remains is to decide how, from the results of one particular Monte Carlo analysis, the variance σ^2 (or its positive root σ, the standard deviation) can be computed. Once the required level of confidence α has been selected, the corresponding value of L/σ can be read from the table and multiplied by the computed value of σ to obtain L, which is half the size of the confidence interval. The most commonly used value of α is 0.95 or 95% which results in a value for L of 2σ.

It can be shown that the variance σ^2 of the pdf $\theta(\hat{Y})$ is related to the true value of yield Y and the number of random circuits (N) sampled in the Monte Carlo analysis by

$$\sigma^2 = Y(1 - Y)/N \tag{4.6}$$

or

$$\sigma = \sqrt{[(Y(1 - Y)/N]} \tag{4.7}$$

---

* The exact form of normalized Gaussian tables varies, although it is easy to convert from one form to another. Here we have arranged the table to suit our purpose which is to find L for our particular choice of α, given σ.

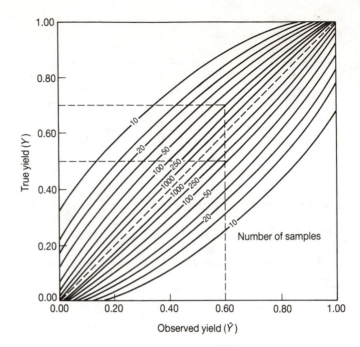

Figure 4.12
Nomogram for calculating 95%
confidence interval for yield
estimation.

Unfortunately, calculation of σ requires knowledge of the true yield which, of course, we do not know; this is why we are carrying out the Monte Carlo analysis! However, an adequate approximation to σ can be obtained from Equation 4.7 by substituting, for Y, the estimate y' of this quantity. Having obtained σ from Equation 4.7 in this way we simply multiply it by the appropriate value of L/σ obtained from Table 4.1 and thus construct the confidence interval $y' \pm 2L$. An even simpler method of obtaining the confidence interval is to use a nomogram such as the one shown in Figure 4.12. This nomogram, unlike the previous table, only applies to one value of the confidence interval α; in the case of Figure 4.12 the selected value of α is 0.95. One pair of curves applies to a particular sample size. To obtain the confidence interval for a sample size of (say) 100, one first locates the observed value (y') of yield on the horizontal axis. Then one reads off the lower and upper limits of the confidence interval on the vertical axis from the two curves corresponding to $N = 100$. In Figure 4.12 such a calculation has been carried out for an observed yield of 60% obtained from a 100 sample Monte Carlo analysis and gives a 95% confidence level of 60% $\pm$ 9.7%.

The interpretation to be placed on confidence intervals is as follows. If each Monte Carlo experiment to estimate yield were to be repeated many times then, on average, 95% of the confidence intervals so constructed would embrace the true value of yield.

For a practical illustration of this interpretation we consider the

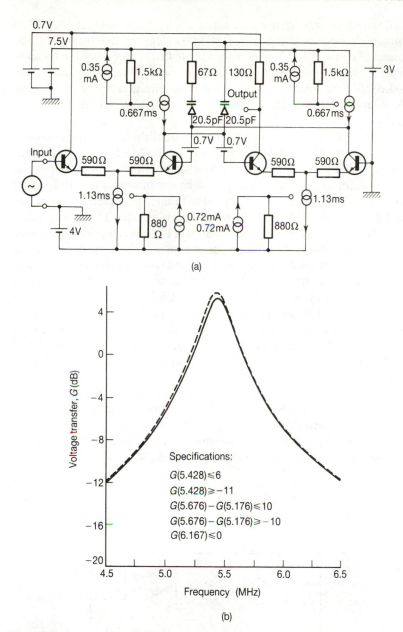

(a)

(b)

Figure 4.13
Gyrator resonator circuit with nominal resistors and capacitors.

integrated circuit shown in Figure 4.13 along with its nominal performance and specifications. One hundred separate Monte Carlo analyses, each in turn involving 100 random sample circuits, were carried out. Figure 4.8 shows a histogram of the estimates of yield that were obtained, and with each yield estimate a 95% confidence interval was constructed. Quite separately and independently another Monte

Carlo analysis was carried out using a very large number (10 000) of sample circuits. In the absence of knowledge of the true yield, the results of this second confirmatory Monte Carlo analysis were taken to provide an answer very close to the true yield. When compared to the 100 confidence intervals constructed from the first set of 100 Monte Carlo analyses, 92 of the confidence intervals were found to embrace the true yield.

It is certainly vital to be able to establish, from a single Monte Carlo analysis, the confidence limits for a selected degree of confidence, but the discussion presented above also allows three important properties of Monte Carlo analysis to be established. First, the greater the required confidence in the yield estimate, the wider must be the confidence limits. Table 4.1 shows that if one wishes to be 99% rather than 95% confident that the true yield will lie between particular limits, then these limits must be 50% wider (from $y' \pm 2\sigma$ to $y' \pm 3\sigma$) for the same number of random circuits. A second consequence of Expression 4.7 is that the size of the confidence interval is inversely proportional to the *square root* of the number of Monte Carlo samples. Thus, having carried out a 100-sample Monte Carlo analysis and obtained an estimated yield with a particular confidence range, it would be necessary to carry out a 400-sample Monte Carlo analysis in order to be equally confident that the true yield lay within a range only *half* as wide.

But it is the third property that can arguably be said to be of the greatest significance. We note from Equation 4.6 that the standard deviation σ (which determines the uncertainty in the yield estimate), is *independent* of the number of parameters associated with the circuit. In this respect – as mentioned in Section 4.2.2 (see Dimensional

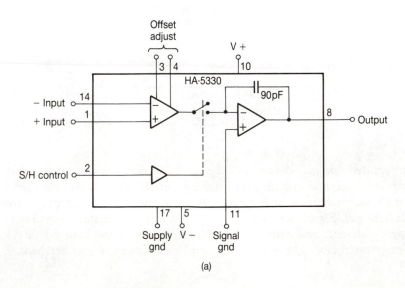

Figure 4.14
(a) Monolithic 'sample and hold' configuration, d.c., a.c and transient Monte Carlo performed on entire 200 + transistor circuit.
(b) Results of Monte Carlo analysis performed on the bipolar 'sample and hold' circuit example.

(a)

Results of
Monte Carlo analysis

Results of measurements
on physical devices

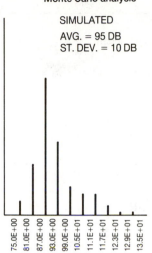

SIMULATED
AVG. = 95 DB
ST. DEV. = 10 DB

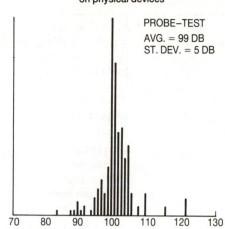

PROBE–TEST
AVG. = 99 DB
ST. DEV. = 5 DB

Cell bounds

Common-mode rejection ratio (dB)

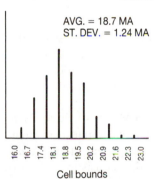

AVG. = 18.7 MA
ST. DEV. = 1.24 MA

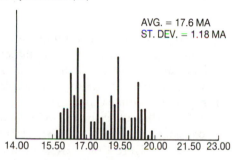

AVG. = 17.6 MA
ST. DEV. = 1.18 MA

Cell bounds

Power supply current

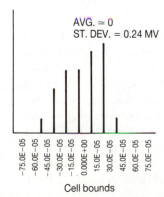

AVG. ≃ 0
ST. DEV. = 0.24 MV

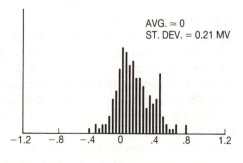

AVG. ≃ 0
ST. DEV. = 0.21 MV

Cell bounds

Offset voltage

(b)

Table 4.2 Performance functions considered for sample-and-hold example (Figure 4.14).

(1) d.c.	Input offset voltage
	Input offset voltage temperature coefficient
	Input offset current
	Input bias current
	Common mode rejection ratio
	Power supply rejection ratio
	Large signal d.c. voltage gain
(2) a.c.	Gain-bandwidth product
	Gain crossover frequency
	Gain margin
	Phase margin
	1 kHz spot noise
	Total integrated noise (1 Hz to 1 GHz)
(3) Transient	Small signal overshoot
	Small signal rise time
	Large signal slew rate
	Large signal settling time

independence) – it is superior to deterministic techniques when the number of components in a circuit exceeds about 5. Experimental evidence also suggests that this same dimensional independence may hold with regard to the derivation of tolerance design information from the results of a Monte Carlo analysis.

To further illustrate the main features of a Monte Carlo analysis we consider the results obtained from a practical circuit example. The circuit (Figure 4.14(a)) is a monolithic sample-and-hold circuit fabricated in a bipolar process and basically consisting of two operational amplifiers, a switch and a buffer. For clarity only an outline circuit schematic is shown in Figure 4.14(a), whereas the detailed circuit contains over 200 transistors. A Monte Carlo analysis employing 100 samples was carried out with the circuit in its 'sample' mode. Table 4.2 contains a list of the performance functions that were evaluated. It should be noted that they cover all three forms of analysis, d.c., a.c. and transient. Figure 4.14(b) also shows two sets of histograms. For three performance functions, the histograms on the left were obtained from the 100-sample Monte Carlo analysis, while those on the right were obtained by measurements on physical devices taken from one fabrication lot. It can be seen that the differences between the averages and standard deviations are small. Moreover, they are explicable in terms of the nature of probabilistic experiments and the fact that the measured results are from one fabrication lot only. Figure 4.15 shows another set of results which relate to the transient analysis

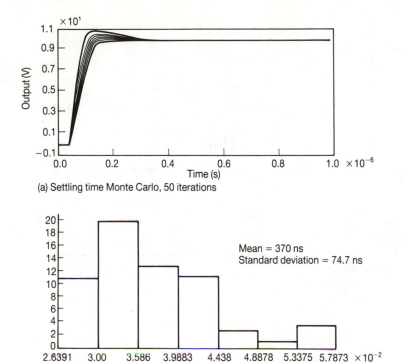

(a) Settling time Monte Carlo, 50 iterations

Mean = 370 ns
Standard deviation = 74.7 ns

(b) Settling time distribution (ns)

Figure 4.15
(a) Output voltage versus time for all 100 random circuit examples.
(b) Histogram of large-signal settling time.

computation of settling time*. In Figure 4.15(a) the entire curve of output voltage against time for each of the 100 sample circuits is plotted on the same diagram. Figure 4.15(b) shows the histogram of settling time, obtained from Monte Carlo analysis. The analysis was carried out on a Harris H800 computer and the CPU time required was 5 h for the d.c. results, 3 h for the a.c. results and 10 h for the transient results.

An example of Monte Carlo analysis applied to a digital circuit is provided by a flip-flop cell fabricated in CMOS technology (Figure 4.16). The circuit designer was mainly interested in the variation of time delay resulting from the statistical variation in device parameters as a function of the load capacitance C_1 whose value was taken to be a deterministic quantity, not subject to statistical variation. Whereas the previous example presented the Monte Carlo analysis results in the form of histograms, in this example two different forms of displaying these results are illustrated. Figure 4.17 shows scatter plots of time delay against four critical MOS device parameters. These are,

* Settling time is defined as the time taken for one output to reach within a certain percentage (often 1%) of its final value.

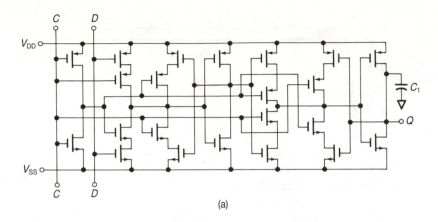

(a)

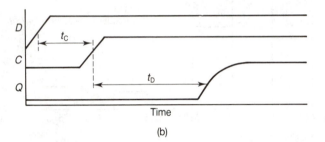

(b)

Figure 4.16
(a) An example of Monte Carlo analysis applied to a digital circuit: a CMOS flip-flop cell.   (b) Typical shape of input and output waveforms for the CMOS flip-flop. C is the input clock, D is data and Q the output.

for the PMOS(NMOS) transistors, the transistor current gains $K_{PP}(K_{PN})$, threshold voltages $V_P(V_N)$, and T_{OX}, the thickness of the gate oxide. Also quoted next to each scatter plot is the correlation between the time delay and the relevant parameter. These results indicate that the time delay is strongly correlated to the PMOS and NMOS current gains whereas it is weakly correlated to their threshold voltages and the thickness of the gate oxide. Another set of results, shown in Figure 4.18, indicates the statistical variation of time delay as a function of load capacitor C_1. The diagram basically shows the mean value of standard deviation of the time delay for three different values of load capacitance C_1. This example required 2 h of CPU time on the Harris H800 computer.

4.3   Practical considerations

We now discuss two issues of vital importance in the application of the Monte Carlo method, which were touched upon only briefly. First, it has been assumed that component parameter distributions are known, but no advice has been given as to how they are obtained and what

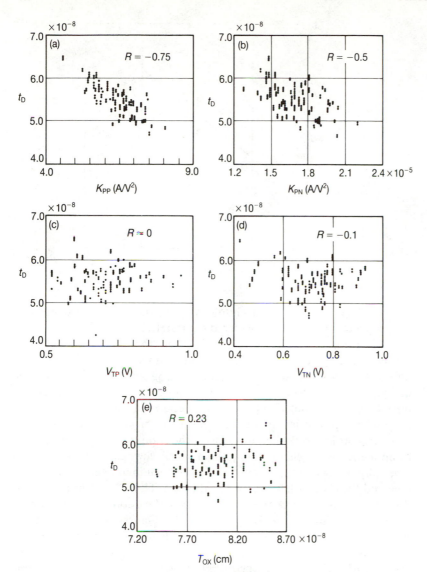

Figure 4.17
Scatterplots of flip-flop time delay t_D versus critical MOS device parameters.

form they take. Second, the underlying mechanism of the pseudo-random generation of points in parameter space has not been explained.

4.3.1 Statistical modelling

A Monte Carlo analysis can only be carried out if the designer has specified the statistical distributions in addition to the nominal values of the component parameters. Since the choice of a statistical model is

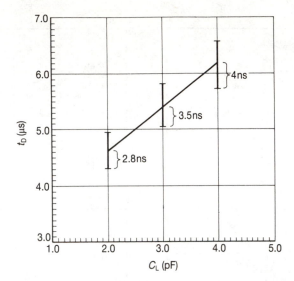

Figure 4.18

Flip-flop time delay mean value $\overline{t_D}$ and standard deviation as function of load capacitor C_L.

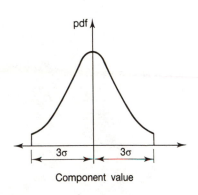

Figure 4.19

Quoted tolerances may be taken to be $\pm\sigma$ points of a Gaussian distribution, e.g. for carbon resistors.

sharply dependent on the technology used, the case of discrete and integrated circuits will be considered separately.

For discrete components it is often the case that statistical models are not provided by the component manufacturer. In this case, unless a very expensive measurement programme is justified, the designer will have to make a judicious choice of the statistical models on the basis of limited information. Becker and Jensen (1977) give suggestions appropriate to this situation, some of which we repeat here.

According to Becker and Jensen, the component pdfs for discrete components are generally Gaussian, with few exceptions. This is especially true when the components have been manufactured for some time and the process has become established. Thus, for carbon resistors (Figure 4.19), a practical expedient is to take the quoted tolerance limits as corresponding to the 3σ points of the Gaussian distribution. A radically different situation arises if a manufacturer selects components from a batch to provide different grades of component (at different prices) with different tolerance limits (Figure 4.20). While the high-precision components in this example could be assumed to have a uniform distribution, the middle and lower grade components would have to be described by bimodal distributions.

With an integrated circuit, all the components are manufactured simultaneously in a series of processing steps. Not surprisingly, therefore, the parameters of the various components within the circuit are highly correlated. Typically, and principally by means of 'test chips' placed on wafers, an integrated circuit manufacturer will obtain a great deal of information concerning the likely variation of various parameters (e.g. current gain, resistivity) within a chip, between chips

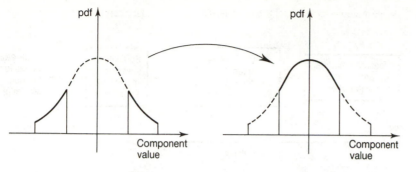

Figure 4.20
Component pdfs resulting from the manufacturer having removed the fine tolerance components for sale at a higher price.

on the same wafer, and over a batch of wafers. Typical variations of current gain for NPN transistors over a slice (wafer) are shown in Figure 4.21. Such data (see Chapter 6 for more extensive discussion) is stored and made available for Monte Carlo tolerance analysis and design. In this case the designer does not have to be concerned with the acquisition of statistical models for his components. The need to characterize parameter spreads in this way, and to characterize them accurately, is clear when one recalls that integrated circuit designers frequently *exploit* parameter correlation to achieve good circuit performance.

Rankin (1982) has described two approaches to the statistical modelling of integrated circuits. The first approach is based on

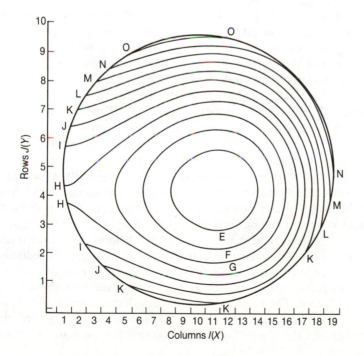

Figure 4.21
NPN current gain. Beta at 0.35 mA. Third-order surface fitted to 116 points after four outliers have been deleted. Average = 77.439. Initial σ = 7.2652 = 9.3818 pct. Surface σ = 3.7136 = 4.7955 pct. E = $-$ 6.000%, F = $-$ 4.000%, G = $-$ 2.000%, H = average, I = 2.000%, J = 4.000%, K = 6.000%, L = 8.000%, M = 10.00%, N = 12.00%, O = 14.00%.

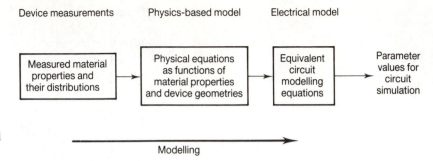

Device measurements   Physics-based model   Electrical model

| Measured material properties and their distributions | Physical equations as functions of material properties and device geometries | Equivalent circuit modelling equations | Parameter values for circuit simulation |

Modelling

Figure 4.22

The statistical modelling of integrated circuits based on physical models.

characterizing the distributions and correlations of the *electrical* parameters. To obtain this information special wafers have to be fabricated in the process in question, each wafer containing a large number of devices of all allowable sizes and geometries. By means of measurements on these devices a large database is built up from which the component parameter distributions and correlations can be extracted using well established statistical techniques. An advantage of this approach is that electrical models are then directly available. Unfortunately, it can be expensive in view of the large number of devices that require testing. A further serious disadvantage is that the fabrication process may change slightly from time to time, so that entirely new measurements are needed to accommodate the effect of these changes.

A second approach to the statistical modelling of integrated circuits is based on physical models. The basic principle is that of relating the parameters of a device's *electrical* model to the *material* properties of the chip on which the device is fabricated (Figure 4.22). The overall approach is indicated in Figure 4.23. The relations between the electrical and material properties, one for each type (not size) of device, are called the modelling equations. Thus, to take the first example of Figure 4.23, a resistor is represented by an electrical equivalent circuit comprising a resistance and two capacitances. The values of the resistance and the capacitances can be computed from knowledge of three material properties (sheet resistivity, over-etch and zero-voltage collector-base capacitance) as well as the dimensions of the resistor. The derivation of the modelling equations is certainly a complex task, involving considerable research by device physicists and process engineers, and is achieved with the help of special chips (called process control modules) included within each wafer. Measurements carried out on the devices contained within the modules allow the material properties, as well as their variation over a number of wafers, to be deduced. The resulting advantage is that when the chip designer wishes to investigate the effect of process variations on circuit performance, the Monte Carlo analysis facility can *automatically* access

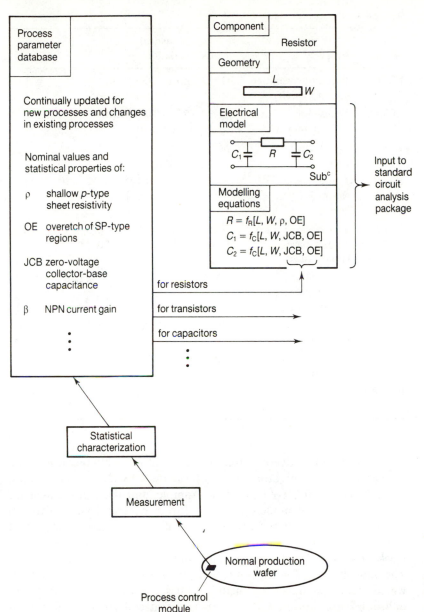

Figure 4.23
The use of process control modules to maintain a database of process parameters from which the values of electrical model parameters can be computed.

data concerning the distributions of the material properties of the current fabrication process. Slight changes in the process do *not* affect the modelling equations; they *do* affect the material properties and their distributions, but this information is automatically and continuously captured from the process control modules.

Table 4.3 Process variables.

Name	Property	Units	Statistical variation
RSBE	Base-under-emitter sheet resistivity	Ω^2	norm (3000, 850)
BETA	Transistor current gain	—	$5.2 + 0.0216\ \text{RSBE}$ $+ \text{norm}\,(0, 5)$
TAUO	Base transit time	ps	$710 - 0.0725\ \text{RSBE}$ $+ \text{norm}\,(0, 30)$
IO	Emitter-base saturation current	A/m^2 A/m^2	$1.43 \times 10^{-10} - 8.52 \times 10^{-8}\ \text{RSBE}$ $+ \text{norm}\,(0, 4 \times 10^{-8})$
CBEJ	Zero-voltage emitter-base capacitance	F/m^2 F/m^2	$11.22 \times 10^{-4} - 9.75 \times 10^{-8}\ \text{RSBE}$ $+ \text{norm}\,(0, 5 \times 10^{-8})$
CBCJ	Zero-voltage collector-base capacitance	F/m^2	$\text{norm}\,(2.165 \times 10^{-4}, 1.08 \times 10^{-5})$
RSP	Resistor sheet resistivity	Ω^2	norm (200, 10)
RDW	Lateral over-etch of resistors	μm	norm (0, 1)
EBDW	Lateral over-etch of emitter regions	μm	norm (0, 1)

As an illustrative example of statistical modelling, consider the bipolar integrated gyrator circuit of Figure 4.13 (Soin and Rankin, 1985). This circuit served as a test example, and therefore somewhat idealized statistical models were employed; nevertheless, it provides a reasonable illustration. The electrical models for each component were produced by geometrical scaling rules from the lateral mask dimensions of the components and nine basic process variables. The models used for the four transistors were Ebers-Moll models which were extended by including base resistances, emitter-base diffusion capacitances, and emitter-base and collector-base depletion capacitances. Parasitic capacitances determined by the d.c. bias conditions and component lateral dimensions were added to the ten resistor models using the collector-base capacitance laws. Series resistances in the two resonating capacitors formed from reverse emitter-base and collector-base junctions were also generated. Consequently, the electrical properties of different components tracked as each process variable was altered.

The nine chosen process variables are listed in Table 4.3, where norm(x, y) indicates values drawn from a Gaussian distribution with mean value x and standard deviation y. To simulate the correlation between properties found in an integrated circuit the variables BETA, TAU, IO and CBEJ were generated from regression equations in the variable RSBE, and therefore were mutually correlated. The attention of the interested reader is drawn to a number of statistical characterization methods recently appearing in the literature (Inohira *et al.*, 1985; Ito *et al.*, 1983).

4.3.2 Generation of pseudo-random component values

True Monte Carlo analysis requires the *random* generation of points in parameter space. In practice, these points are generated within the computer by an algorithm which, because it cannot achieve a truly random result, is referred to as a *pseudo-random* process. Many such algorithms exist: here we shall describe only the salient features of this type of algorithm by means of an illustrative example.

The basis of a general pseudo-random number generator capable of a wide variety of probability distributions is a method of generating numbers from a uniform distribution in the range 0 to 1. The basic procedure used is a (necessarily deterministic) recurrence relationship, often of the form

$$x_i = (ax_{i-1} + c)_{\text{modulo } m} \tag{4.8}$$

which is referred to as a *mixed congruential generator*. The multiplier a and the increment c are non-negative integers in the range 0 to $m - 1$. The modulus m is usually a large power of 2. The notation $(\Phi)_{\text{modulo } m}$ simply means

$$(\Phi)_{\text{modulo } m} = (\Phi) - m \times \text{int} \, (\Phi/m)$$

where int (Φ/m) means the integer part of the quotient (Φ/m).

Expression 4.8 gives an integer in the range 0 to m, which is then divided by m to give a value between 0 and 1, i.e. $y_i = x_i/m$. As an example we take $a = 3$, $c = 5$, $m = 8$ and an initial value of 2 for x; i.e. $x_1 = 2$. Then Equation 4.8 will give the sequence $x_1 = 2$, $x_2 = 3$, $x_3 = 6$, $x_4 = 7$, $x_5 = 2$. We stop at x_5 because $x_5 = x_1$ and the recurrence relationship will ensure that the sequence repeats itself. Clearly, all such sequences will repeat themselves with a period less than or equal to m. In the example above the period was only four whereas the modulus m was eight. To ensure a large period, m is therefore usually chosen to be a large integer, for example 2^{31} for a computer with a world length of 32.

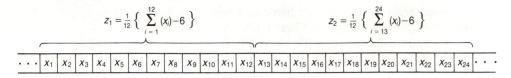

$$z_1 = \frac{1}{12}\left\{ \sum_{i=1}^{12}(x_i)-6 \right\} \qquad\qquad z_2 = \frac{1}{12}\left\{ \sum_{i=13}^{24}(x_i)-6 \right\}$$

$\cdots$ | x_1 | x_2 | x_3 | x_4 | x_5 | x_6 | x_7 | x_8 | x_9 | x_{10} | x_{11} | x_{12} | x_{13} | x_{14} | x_{15} | x_{16} | x_{17} | x_{18} | x_{19} | x_{20} | x_{21} | x_{22} | x_{23} | x_{24} | $\cdots$

Figure 4.24

Scheme for the generation of a sequence of numbers (y_i) obeying a Gaussian distribution from another sequence (x_i) distributed uniformly.

In the course of a Monte Carlo analysis, relation 4.6 will be invoked many times. Clearly, if one starts with a particular starting value x_1 (28 041, say), then the rest of the sequence $x_2, \ldots, x_N$ is determined. Thus, by setting the starting value (called the random seed) of the sequence x to a previous value, an entire sequence of numbers can be reproduced. Such a facility can be useful in certain applications such as 'correlated sampling' discussed in Chapter 7.

The reader may question whether a basically deterministic procedure such as Equation 4.8 can generate a sequence of values that can be considered, even approximately, to be random. The explanation lies in the fact that large sequences of numbers generated by such deterministic procedures have statistical properties similar to those that would be expected of sequences of values obtained from truly random physical processes.

Therefore, for purposes of Monte Carlo or any other computation requiring random values, the results are unaffected by the fact that the values were not obtained from a truly random process. The main reasons for the use of such pseudo-random generators instead of a random physical process are those of convenience and speed and the fact that pseudo-random number sequences are reproducible. The reader can be assured that, before implementation, an algorithm for pseudo-random number generation will have been subjected to a battery of appropriate statistical tests to ensure that the sequences generated are satisfactory.

Monte Carlo analyses must, of course, be capable of handling *any* probability distribution of the component parameters, and must therefore contain some means of generating numbers whose distribution is the one required. To do this, a sequence of numbers approximately uniformly distributed between 0 and 1 is first generated by an algorithm such as the one just described. Then, a suitable transformation is employed to generate a new sequence having the required distribution. One such transformation suited to the generation of values according to a Gaussian distribution will be used to illustrate the principle.

One algorithm (Figure 4.24) for the transformation of a pseudo-random uniform distribution to one that is Gaussian is

$$z_i = \{x_{i1} + x_{i2} + x_{i3} + \ldots + x_{in} - n/2\}/n \qquad \text{for } 0 \leqslant x_i \leqslant 1$$

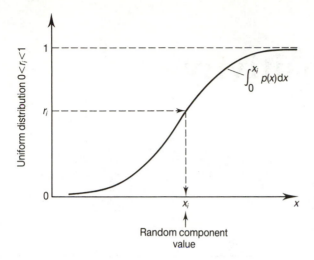

Figure 4.25
A cumulative distribution function.

where z_i is a value from the required Gaussian distribution and x_{i1} to x_{in} are n successive values of a uniform distribution. According to the Central Limit Theorem (Larson, 1969) the distribution of z_i will approach Gaussian for sufficiently large n: typically a value of $n = 12$ is satisfactory. The term $n/2$ is introduced to ensure that the mean value of the z_i is zero. The variance of the z_i can be shown to be equal to $n\sigma_x^2$, where σ_x^2 is the variance of the x_i. To obtain a Gaussian distribution having a non-zero mean and a particular variance only simple translation and scaling are involved.

A more general method and one that may be used for many different types of component pdf is based on their cumulative distribution function. The cdf $F(x)$ of a pdf $\Phi(x)$ is defined as

$$F(x) = \int_{-\infty}^{x} \Phi(t)\, dt$$

In other words, it is the integral of the pdf. For use of the method see Figure 4.25. A pseudo-random value in the range 0 to 1 is first generated, usually by the congruential method described earlier. This is first marked off on the vertical axis and its corresponding component value on the horizontal axis is read. In practice the cdfs are often available in piecewise linear form or as tables. The justification for this procedure is discussed in several references (e.g. Rubinstein, 1981).

CHAPTER 5

Tolerance Sensitivity

OBJECTIVES

The data generated by a Monte Carlo analysis is capable of providing far more than just an estimate of yield and its associated confidence limits. With little additional computational effort, insight can also be obtained into the effect on the unwanted consequences of component tolerances of the adjustment of nominal parameter values, parameter tolerances and performance specifications – how the likelihood of a circuit passing the specifications is related to the actual nominal values of individual components. In other words, the circuit designer has some idea of how the adjustment of a component's nominal value, or the choice of a new tolerance, will affect the manufacturing yield. Such information is generally referred to as **tolerance sensitivity** and forms a link between tolerance analysis and tolerance design. The chapter also shows how the results of a Monte Carlo analysis can illuminate any trade-off that might be possible between individual specifications and the manufacturing yield. The success with which the tolerance sensitivity information can be utilized depends, however, on the skill and experience of the designer. Finally it is shown how, from the results of a Monte Carlo analysis, the sensitivity of manufacturing yield to changes in nominal component values and tolerances can be determined.

5.1   Introduction

We have seen in Chapter 4 how the Monte Carlo technique can be used to estimate the manufacturing yield of a mass-produced circuit, as well as establish the limits within which the actual yield can be expected to lie with a known level of confidence.

Whether this estimated yield is 100% or less, it is profitable for the designer to enquire how the yield might be affected by, for example, changes in nominal component values, changes in parameter tolerances, or changes in the customer's specifications upon circuit performance. If changes in a particular specification have, for example, no effect on the estimated yield, then it may be appropriate to tighten that specification so that an improved product can be offered for sale. The topic of tolerance sensitivity embraces these and similar questions.

Fortunately, a substantial amount of information concerning tolerance sensitivity is contained within the results of a Monte Carlo analysis and can be obtained with little additional computational effort. In this chapter we see how this information can be extracted and usefully applied.

5.2   Parameter histograms

Let us assume that a Monte Carlo analysis has been carried out for a particular circuit, and that a database (Figure 5.1) has been created which stores, for each sample circuit, the parameter values and the information (1 or 0) concerning whether the sample's performance was classified as pass or fail. Note that, in this instance, no information is stored in this database concerning the circuit's performance other than whether it passed or failed the specifications. From this database, and for one selected parameter p_i, two histograms can now be plotted showing the densities of passes and fails to a base of the parameter value (Figure 5.2). For convenience in the initial discussion, the histograms are shown as continuous curves even though, in practice, the p_i axis will be divided into a number of class intervals and the histograms will

Monte Carlo sample	$R\,(\Omega)$	$C\,(\mu F)$	Pass (1) or fail (0)
1	1037	0.00104	1
2	1074	0.00092	0
3	925	0.00109	1
⋮	⋮	⋮	⋮

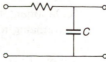

Figure 5.1
A circuit and the database generated by a Monte Carlo analysis.

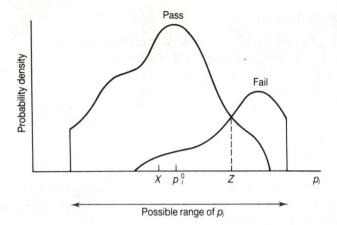

Figure 5.2
A pass–fail histogram associated
with one parameter.

therefore be discrete in nature. The histograms are called pass–fail
histograms.

An immediate reaction to the pair of histograms shown in
Figure 5.2 might be that a greater proportion of pass circuits is likely to
be manufactured if the nominal value of p_i is reduced somewhat (though
with the same tolerance) to a value in the vicinity of the point X. This
conclusion must be applied with caution, however, because we have no
information whatsoever concerning how the pass and fail histograms
would appear beyond the limits to the current sampling range. Another
common reaction to the histograms of Figure 5.2 is to suggest that, if
both upper and lower limits to p_i can be adjusted independently, the
yield might be increased if the upper limit is adjusted to the value Z,
quite empirically chosen as the value of p_i for which the histograms
intersect.

Aware of the design guidelines that pass–fail histograms may
provide, Norman Elias (a pioneer of the statistical sampling approach
to tolerance design) suggested that some measure of the overlap
between the pass and fail histograms associated with a given parameter
$p_i \ldots$

> is a direct measure of the influence of parameter p_i on circuit
> yield ... The more sensitive the circuit yield is to any particular
> component, the less should be the overlap between values of that
> parameter in passing and failing circuits. (Elias, 1975)

Accordingly, Elias proposed the quantity M_i, where

$$M_i \triangleq \int_{-\infty}^{+\infty} |p_{\text{pass}}(p_i) - p_{\text{fail}}(p_i)| \, \mathrm{d}p_i \tag{5.1}$$

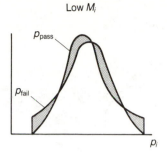

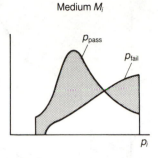

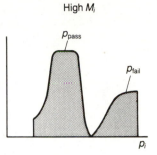

Figure 5.3
Pass–fail histograms exhibiting different degrees of overlap.

as a scalar measure of the difference between the two histograms, where p_{pass} and p_{fail} are the probability densities associated with the pass and fail histograms for parameter p_i. The value of M can vary between 0 (total overlap) and 2 (no overlap). Once having carried out a Monte Carlo analysis it is a relatively straightforward task to compute, for every parameter within the circuit, the corresponding value of the overlap M. The designer's attention can then be drawn to those components with small overlap. An example is shown in Figure 5.3.

5.3   Simple tolerance design

If a particular parameter exhibits a large M value (i.e. a small degree of overlap), what steps should be taken by the designer? Elias describes a number of ways of exploiting the information contained within the histograms. For example, he describes a method of tolerance assignment (though, to be precise, it can only lead to a *tightening* of tolerance limits) in which the upper and lower limits on p_i are adjusted, where appropriate, to the values for which $p_{pass} = p_{fail}$; in other words, the parameter values for which the histograms intersect. Two examples showing old and new ranges for a parameter are shown in Figure 5.4. Although Elias used a statistically based proof that such a scheme would always increase the manufacturing yield, that proof assumed that the limits on only one parameter were adjusted, all other parameter limits remaining fixed. There is therefore no assurance that the application of the technique simultaneously to more than one parameter will lead to a yield increase. Indeed, simultaneous application of the tolerance tightening algorithm to a number of parameters may result in 'overkill' in the sense that the increased cost associated with the new limits is disproportionately higher than the savings that accrue from the yield increase.

    If either the sensitivity measure M is computed, or the tolerance tightening scheme is applied, it would still be advisable for the designer to have sight of the histograms, since rules which appear reasonable

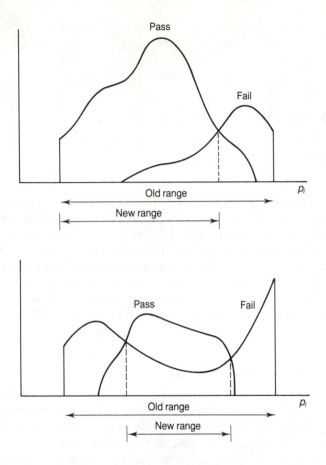

Figure 5.4
Illustrating tolerance tightening based on pass–fail histograms.

when viewed in the ideal context of continuous and unimodal histograms may not be as easily applied, if at all, in practical situations involving a finite number of Monte Carlo samples and a distribution which is not unimodal.

5.4   Selection of components before circuit manufacture

Elias (1979) has also shown how the concepts discussed above can be applied to the selection of components before circuit manufacture, with a view to increasing the manufacturing yield. The scheme is outlined in Figure 5.5 for the simple case of a four-parameter circuit. First, a Monte Carlo analysis is carried out, and the *M*-factors computed for each parameter. Let us suppose that one of these four parameters has a much higher *M* value than the others. The pass–fail histogram for this parameter is then examined, and a test limit established which, if satisfied, will lead to an acceptable value of manufacturing yield. Then,

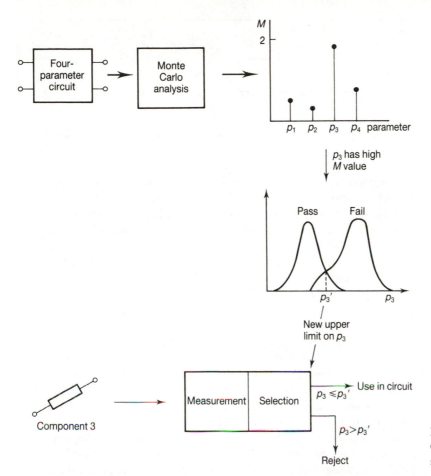

Figure 5.5
Component selection by prior measurement of critical components.

in the circuit manufacturing process, the component whose parameter has been so characterized is first measured; only if its parameter value lies within the test limits is it considered acceptable and connected within the manufactured circuit. In this way the cost of pre-production parameter measurement may be more than offset by the increased manufacturing yield. For components with low M values such pre-manufacture measurement may not be economic in view of the small probability that a deviation in their value will cause the circuit to fail its specifications.

5.5   Specification sensitivity

The designer may wish to know the effect on the estimated manufacturing yield $\hat{Y}$ of a change in one of the specifications on circuit performance, to judge the nature of any trade-off that might be possible.

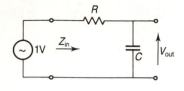

Monte Carlo sample	R (Ω)	C (μF)	$\|V_{out}\|$ (V)	$\angle\, V_{out}$ (degrees)	$\|Z_{in}\|$ (Ω)	$\angle\, Z_{in}$ (degrees)	pass (1) fail (0)
1	991	0.00098	0.53	37	1401	37	1
2	1017	0.00104	0.57	42	1376	42	0
⋮	⋮	⋮	⋮	⋮	⋮	⋮	⋮

Figure 5.6
The form of a database built up in the course of a Monte Carlo analysis.

For example, it could be useful to know that, for a 0.2 dB change in the upper limit specification on filter loss at 900 Hz, an extra 4% of yield can be achieved: in this case the customer might be approached to see if, and to what extent, the trade-off should be exploited. The change in specification might be small, in which case it would be the differential sensitivity $(\partial Y/\partial F_J^+)$ of yield with respect to change in the upper $(+)$ limit of the specification on circuit performance F_J that would need to be computed. On the other hand it may also be useful for the designer to be able, dynamically and interactively, to explore the effect of large variations in a specification.

Before proceeding to show how specification sensitivity information can be generated it must be clearly stated that, to do so, it is necessary to store the *numerical values* of the circuit performance calculated in the course of the Monte Carlo analysis. In other words, the database produced as the result of the Monte Carlo analysis must have the form shown in Figure 5.6. Normally, of course, after the analysis of a sample circuit, its performance is checked against specification and only a pass or fail recorded, after which all other performance details are discarded: the database of Figure 5.1 was generated in this way. To generate specification sensitivity information, however, full performance details must be retained, as shown in Figure 5.6. The amount of storage needed to retain performance details will, in general, not be negligible, and there is therefore a need for the user to be aware at the start of a Monte Carlo analysis that he must declare whether specification sensitivity will later be of interest. It may well be possible, of course, that as the design proceeds it becomes clear that specification sensitivity will only be of interest at one or two performance limits, in which case the storage burden could be eased.

To discuss the topic of specification sensitivity, and thereby derive an algorithm for its calculation, we first assume that a Monte Carlo analysis has been carried out for the circuit and that specifications (i.e. limits of acceptability) have been placed on the numerical values of three circuit properties denoted F_1, F_2 and F_3. These could, for example, be different circuit properties at the same frequency or the same property at three different frequencies. First, from the results of

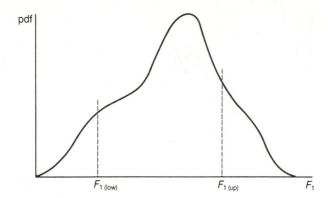

Figure 5.7
Distribution of performance F_1.

the Monte Carlo analysis, we construct a curve (Figure 5.7) showing the relative frequency of occurrence of values of F_1 exhibited by the random sample of circuits analysed. For convenience a smooth curve has been shown although, in practice, the finite sample of results will only allow us to plot a histogram with a limited number of class intervals. Note that, until the specifications on F_1 are selected, the curve labelled F_1 in Figure 5.7 contains no yield information whatsoever. Naturally, the area under the F_1 curve is unity, since all samples involved in the Monte Carlo analysis are accounted for. If the upper and lower limits $F_{1(up)}$ and $F_{1(low)}$ on F_1 are now chosen, as shown in the figure, the partial yield can be found: it is the area under the F_1 curve between the limits $F_{1(low)}$ and $F_{1(up)}$.

   We referred above to a *partial* yield because it only applies to the specifications on F_1: it is that proportion of manufactured circuits that meet the specifications on F_1 *irrespective* of whether they meet the other specifications. Similar partial yields could be defined and estimated for any of the other specifications, but any partial yield can only be equal to or greater than the overall manufacturing yield, a measure which simultaneously takes into account all the specifications.

   Now consider the other two responses F_2 and F_3. For each interval on the horizontal F_1-axis we find, by referring to the results of the Monte Carlo analysis (i.e. the database as illustrated in Figure 5.6), how many samples *fail* either the F_2 or the F_3 specification or both. This information is now plotted as the curve $F'_{2,3}$ in Figure 5.8. The $F'_{2,3}$ curve cannot, of course, rise above the F_1 curve because there cannot be more failed samples having F_1 values in a certain class interval than the number of samples in that interval. The $F'_{2,3}$ curve can *touch* the F_1 curve, in which case it would follow that all circuits having an F_1 value in the interval in question failed one or both of the specifications on F_2 and F_3. In general the $F'_{2,3}$ curve will lie below the F_1 curve.

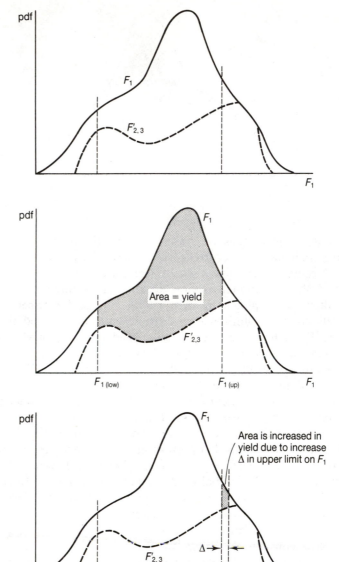

Figure 5.8
Distributions of performance F_1 and the corresponding distribution of F_1 values associated with circuits failing specifications on F_2 and/or F_3.

Figure 5.9
Showing how yield is related to specifications on F_1, F_2 and F_3.

Figure 5.10
Illustration of the sensitivity of yield to changes in the upper specification on F_1.

Before going any further it should be noted that, for any response (e.g. F_1) on which a specification is placed, a plot of the form just discussed (showing the fail curve, e.g. $F'_{2,3}$, relating to the remaining response specifications) can provide the user with useful design information. For example (Figure 5.9), for given limits on F_1, the approximate yield can be assessed: it is the area between the two curves and between the two limits (remember the area under the entire F_1 curve is unity).

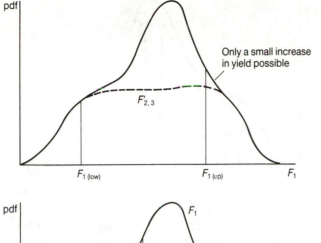

Figure 5.11
Illustrative of a design in which only a small increase in yield can be achieved by adjustment of the upper specification on F_1.

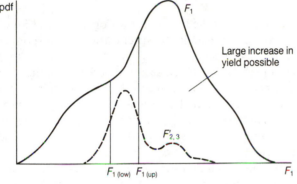

Figure 5.12
Illustrative of a design in which there is considerable potential for an increase in yield by adjustment of the upper specification on F_1.

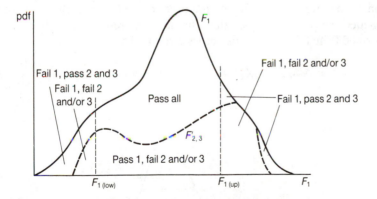

Figure 5.13
The specifications failed for specific choice of the limits on F_1.

Thus (Figure 5.10) the effect of an adjustment of the upper limit on F_1 can be assessed: the yield increases by the area, between the curves, uncovered by the increase Δ in the upper limit on F_1. It follows, therefore, that the situation shown in Figure 5.11 is such that very little increase in yield is gained by increasing the upper limit on F_1, though there is much to be gained in this way for the case shown in Figure 5.12. The illustration of Figure 5.13 may help: it shows the specifications

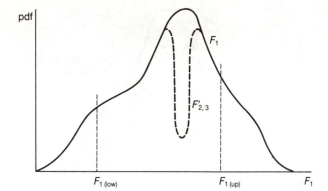

Figure 5.14
The yield is determined principally by specifications on responses other than F_1.

failed for specific choices of the limits on F_1. For the case of Figure 5.14 the yield is determined principally by specifications on responses other than F_1: thus, the specifications on F_1 could be tightened to offer a better product or perhaps be offered as a trade-off against the relaxation of another specification which currently has a detrimental effect on yield. It is seen, therefore, that the designer can make good use of the type of plot we have been discussing: after indicating that specification sensitivity is of interest, the user would only have to indicate the response of interest.

The manner in which an estimate of the specification sensitivity can be computed is easily derived from Figure 5.15. Let the upper limit on F_1 be increased by Δ. The increase in yield is then the area shown shaded, since the samples that this area represents have been transformed from failures to satisfactory circuits. Thus

$$\delta Y = \Delta(M^+_{F_1} - M^+_{F'_{2,3}}) \tag{5.2}$$

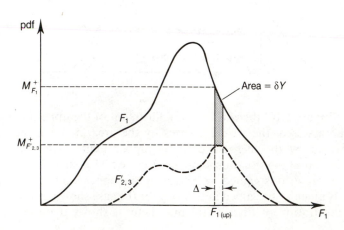

Figure 5.15
Specification sensitivity calculation.

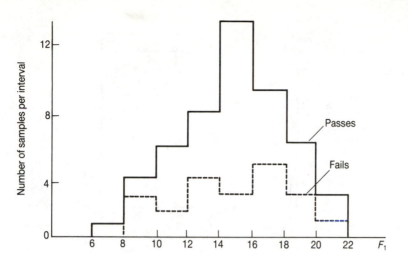

Figure 5.16
Numerical example illustrating the
effect, on yield, of a small change in
the upper specification on F_1. At the
chosen upper limit ($F_1 = 17$), the
number of samples over an interval
of (18−16) in F_1 is nine out of a total
of 50. Thus

$$M_{F_1}^+ = \frac{9}{50} \times \frac{1}{2}$$

Similarly

$$M_{F_{2,3}}^+ = \frac{5}{50} \times \frac{1}{2}$$

So

$$\frac{\delta Y}{\Delta} = \frac{4}{100} = 0.04$$

where δY is the change in yield, $M_{F_1}^+$ is the density of F_1 values and
$M_{F_{2,3}}^+$ the density of fails due to specifications on F_2 or F_3 or both,
both densities being evaluated at the original upper limit on F_1. A
numerical example is given in Figure 5.16.

Thus, for an increase or decrease of
one in F_1, there is an increase or
decrease of 4% in the yield.

As noted earlier, the designer may wish to explore large changes
in (say) the upper limit on a certain response. The manner in which this
can be achieved can be demonstrated using the above numerical
example (see Figure 5.17). For each interval along the F_1 axis the
difference between the two curves gives the yield increase that is
available if F_1 values in that interval are acceptable. Thus, if the lower
limit on F_1 is set at the value 12, then variation of the upper limit from
12 to 22 will result in the yield variation shown in Figure 5.18. In fact
the designer may find a cumulative curve such as the one shown in
Figure 5.19 more useful, since variations in both upper and lower limits
on F_1 can be simultaneously explored.

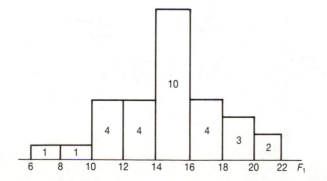

Figure 5.17
Difference between numbers of pass
and fail samples within intervals of
F_1.

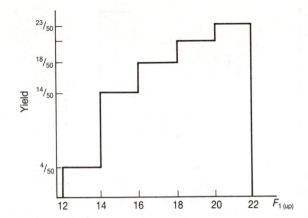

Figure 5.18
Variation of yield with upper specification on F_1, for example of Figure 5.16.

5.6   Yield sensitivity

The most important problem in tolerance design is that of yield maximization. In conventional optimization, methods based on gradients (sensitivities) of the objective function with respect to the design variables are preferred for their greater effectiveness than the non-gradient-based methods. We may therefore reasonably enquire whether Monte Carlo analysis, in addition to providing estimates of such quantities as yield, can also provide the *gradients* of yield with respect to component nominal values and, if applicable, component tolerances. The answer is in the affirmative, as we shall soon see (Batalov *et al.*, 1978). First, however, we present a brief pictorial interpretation of yield gradients.

Figure 5.20 illustrates the case of a simple circuit involving only two components, p_1 and p_2. The manufacturing yield is simply the

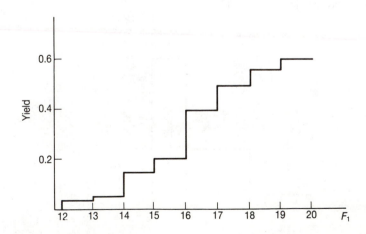

Figure 5.19
Cumulative yield curve.

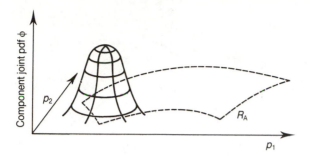

Figure 5.20
The joint pdf and region of acceptability for a two-dimensional parameter space.

volume bounded by the joint component pdf Φ and the region of acceptability R_A which, in this case, is two-dimensional. Consider now an even simpler example involving only one variable component p. The region of acceptability now reduces to a line AB (Figure 5.21). For further simplicity let the component pdf $\phi(p, p^0)$ be bell-shaped and centred about p^0 as illustrated in Figure 5.22. Yield is now the area under this pdf between the limits A and B.

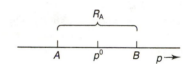

Figure 5.21
A one-dimensional parameter space and region of acceptability.

   Now consider a small positive change Δp^0 in the nominal value, a change that shifts the whole pdf curve along the p-axis. The change in yield ΔY due to such a change Δp^0 in p^0 is indicated as the difference in areas under the two curves between the limits A and B (Figure 5.23). The yield gradient $\partial Y/\partial p^0$ is therefore simply the limit

$$\partial Y/\partial p^0 = \Delta Y/\Delta p^0 \quad \text{as} \quad \Delta p^0 \to 0 \tag{5.3}$$

To proceed further we now return to the estimation of yield gradients via Monte Carlo analysis. First, we define yield as the integral

$$Y = \int_{R_T} \cdots \int [g(P)]\,\phi(P, P^0)\,\mathrm{d}P \tag{5.4}$$

We have expressed the component pdf $\phi(.)$ as a function of both p and p^0 because we are considering changes in its centre p^0 (say) while keeping its basic shape unchanged. As before, $P^0 = p_1^0 p_2^0 p_3^0 \ldots p_K^0$.

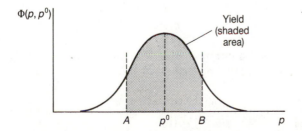

Figure 5.22
Yield of a one-parameter system.

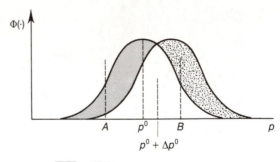

$$\Delta Y = \boxed{} - \boxed{}$$

between A and B

$$\frac{\partial Y}{\partial p^0} = \lim_{\Delta p^0 \to 0} \frac{\Delta Y}{\Delta p^0}$$

Figure 5.23

Pertinent to the consideration of the yield sensitivity of a one-parameter system.

Differentiation of Equation 5.4, and interchanging the order in which the integration and differentiation are performed, then yields

$$\frac{\partial Y}{\partial p_i^0} = \int_{R_T} \cdots \int g(P) \frac{\partial \phi(.)}{\partial p_i^0} \tag{5.5}$$

It will be shown that it is profitable to rewrite Equation 5.5 as

$$\frac{\partial Y}{\partial p_i^0} = \int_{R_T} \cdots \int \left[\frac{g(P)}{\phi(.)} \cdot \frac{\partial \phi(.)}{\partial p_i} \right] \phi(.) \mathrm{d}P \tag{5.6}$$

under the assumption that the pdf $\phi(.)$ is differentiable and non-zero in the region of acceptability.

The reader is reminded that, conventionally, Monte Carlo analysis estimates the value of yield as defined in Equation 5.4 by generating a number N of sets of component values (i.e. $p_1, p_2, p_3, \ldots, p_N$) pseudo-randomly selected from the pdf $\phi(.)$, analysing the resulting circuits to compute the testing function $g(p_i)$ for $i = 1$ to N, and then evaluating the summation

$$\hat{Y} = \frac{1}{N} \sum_{i=1}^{N} g(p_i) \tag{5.7}$$

Comparison of Equations 5.4 and 5.7 indicates that the integral 5.6 could also be estimated by a similar summation:

$$\frac{\partial \hat{Y}}{\partial p_i^0} = \frac{1}{N} \sum_{i=1}^{N} \frac{g(p_i)}{\phi(p_i, p^0)} \frac{\partial \phi(p_i, p^0)}{\partial p_i^0} \tag{5.8}$$

The pseudo-random component values p_i and the corresponding results $g(p_i)$ of the circuit analyses are available from the Monte Carlo analysis performed to compute Equation 5.7. Therefore, the only additional cost is that due to the modest computational effort required to evaluate Equation 5.8. Very similar formulae could be developed for higher order gradients (e.g. $\partial^2 Y/\partial(p_1^0)^2$) and are susceptible to simple computation from the results of the same Monte Carlo analysis. The methods outlined above render the Monte Carlo analysis particularly effective for tolerance design, a property which is exploited by several algorithms of which two are described in Chapter 8.

5.7 Conclusions

We have seen how information can easily be obtained from the results of the Monte Carlo analysis which can indicate on which parameters the tolerances should be adjusted in order to give an increase in yield. This information can also be used to select components before manufacture, with a view to reducing circuit cost by increased yield. Specification sensitivity was also discussed, and it was shown how information from the Monte Carlo analysis can be used to determine trade-offs between circuit yield and the specifications placed on the circuit's performance. Finally, a method of computing yield sensitivity was described.

CHAPTER 6

An Overview of Tolerance Design

OBJECTIVES

We now begin to consider methods for reducing the unwanted effects
of component tolerances, though in a general manner compatible
with providing a context for the three remaining chapters. The means
by which such improvements can be brought about are briefly
reviewed in outline, with separate consideration given to discrete and
integrated circuits. The two principal approaches to tolerance design
are then described: they are known as the deterministic approach and
the statistical exploration approach. Their characteristics are compared, and reasons are given for our subsequent exclusive consideration of the statistical exploration approach. Finally, some criteria for
the evaluation of a tolerance design algorithm are provided.

6.1   Tolerance design

Chapters 3 and 4 have shown how the manufacturing yield of a mass-produced circuit can be estimated. If the yield is unacceptably low, insight concerning the parameters principally responsible can be provided, thereby enabling the circuit designer to suggest desirable adjustments to parameter nominal values and tolerances (Chapter 5). Nevertheless, to complement the designer's skill in exploiting this information, a systematic method of identifying the necessary adjustments would be of considerable value. It is the use of such methods that is often referred to as tolerance design.

The activities of yield maximization and cost minimization have already been described in outline. Figure 6.1, for example, illustrates again the improvement in yield that may be brought about through design centring by a new choice of nominal parameter values while the parameter tolerances are kept fixed. Figure 6.2 is provided to remind us that it may be profitable, after a design has been centred, to leave the nominal values fixed but to adjust the tolerances in order to minimize the cost of each acceptable circuit. However, while these figures provide simple illustrations of two principal components of tolerance design, they also draw attention to a major difficulty: in both figures the boundary of the region of acceptability is shown whereas, in practice, we have little idea of its whereabouts. The principal task of any method of tolerance design is, in fact, to obtain some idea of the whereabouts of R_A.

Before we examine how this might be done we must recognize that the design freedoms implied by Figures 6.1 and 6.2 (the independent and continuous choice of nominal values and tolerances) may not always be available in practice. Indeed, the degrees of design freedom are so technology dependent that we now give separate consideration to the design of discrete and integrated circuits.

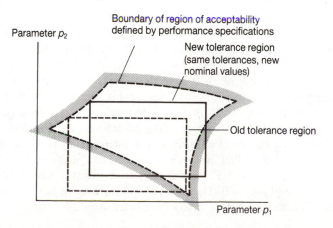

Figure 6.1
One step in the process of design centring, whose objective is the maximization of yield.

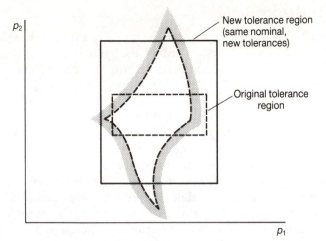

Figure 6.2
Wider tolerances, though leading to a lower manufacturing yield, may nevertheless greatly reduce the total component cost.

6.2   Discrete circuit design

The designer of a discrete component circuit can normally, at least for passive components such as resistors and capacitors, exercise independent choice of the nominal values and tolerances. Thus, a change in the value of the resistance of a particular resistor, or its tolerance, need have no repercussions for the values of other remaining passive components. The tolerance design situation for such passive components is therefore as shown in Figure 6.3 which assumes, for convenience of illustration, a two-dimensional parameter space and a two-dimensional performance space. The tolerance and acceptability regions are identified in both spaces. The transformation from parameter space to performance space is, as we know, nonlinear, and it is usual to employ a circuit analysis package to generate a point-by-point transformation in this direction. By contrast, mapping in the opposite direction – to generate R_A in parameter space given the specifications in performance space – can only be achieved by a process of searching. If the parameter distributions are Gaussian, then the equi-probable contours (i.e. the contours of the joint pdf) are either concentric circles (spheres) for non-correlated components or concentric ellipses (ellipsoids) where there are correlations. The nature and degree of design freedom that is available to the circuit designer is indicated in the figure: it includes *nominal values*, *tolerances* and, though rarely exploited, the *shape* of the parameter distributions.

Often, the choice of nominal component values is restricted to a finite number of so-called 'preferred values'. Thus, if the resistor values are selected from the E12 series, a specific resistor can take on a nominal value of (say) $1.0\,k\Omega$ or $1.2\,k\Omega$, but nothing in between such as $1.15\,k\Omega$. The same kind of restriction often applies to the choice of

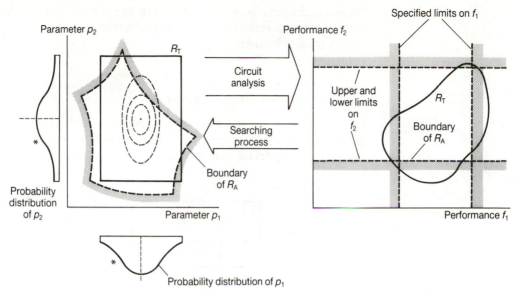

Figure 6.3

The parameter and performance spaces associated with a discrete-component circuit. Asterisks locate the available design freedom, which includes the nominal value, tolerance and distribution of each parameter.

tolerances, which may typically be constrained to values such as 1%, 2%, 5% and 10%.

The characterization of a single component by a single parameter (e.g. a resistor by its resistance) is rarely possible for an active device such as a transistor. For a bipolar transistor a range of parameters for the hybrid-π model, such as the current gain (β), mutual conductance (g_m), output resistance (r_o) and input capacitance (C_π) may well substantially influence circuit performance. Moreover, the values of these parameters are often heavily correlated (e.g. r_o and g_m). Thus, the independent choice of such parameters cannot be contemplated; rather, the designer must examine available devices and select the one that may be optimum for tolerance purposes. The problem is additionally complicated by the dependence of small-signal parameters on the bias condition, since the choice of a new nominal value for a resistor may lead to change in the quiescent condition of one or more active devices, and hence also the small-signal performance of the entire circuit.

With discrete components the main source of uncertainty regarding their characterization is associated with parameter distributions. It is rare for component manufacturers to supply parameter distributions, and it is costly (usually prohibitively so) to carry out in-house measurements to establish a distribution that, it must be noted, would apply only to those components currently in stock; the next batch received from the manufacturer might be characterized by a totally different distribution! It will be seen in the following chapters how the designer, in the face of such uncertainty, can nevertheless make sensible and reasonable assumptions regarding parameter distributions.

6.3   Integrated circuit design

Consideration of the nature of integrated circuit design can be helped by the diagram of Figure 6.4, since it identifies distinct 'layers' of significant variables (shown in boxes) and the transformations between

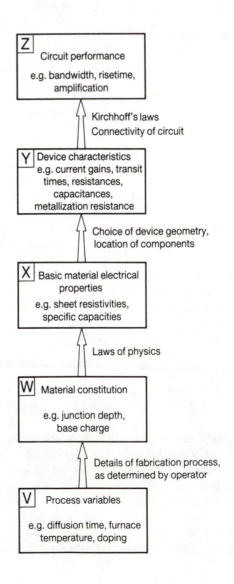

Figure 6.4
The significant variables and transformations associated with integrated circuit fabrication, design and performance.

them. At the lowest level (V) we have variables such as the temperature of a furnace or the time during which impurity diffusion into silicon occurs; the choice of a value for one of these variables does not constrain the choice of the others. In other words, the level V variables are independent. What determines the values of the variables at the next higher level (W) is, in addition to the values of the V variables, the transformation determined by the human operator's choice of the fabrication process. However, since level W variables (e.g. junction depth) typically depend on more than one process variable, correlations are to be found among the W variables. Transformation to the next level (X) is determined purely by the laws of physics; thus, major determinants of sheet resistivity include carrier concentration, impurity levels and the type of charge carrier. Again, some correlation will be exhibited between the X variables. The next transformation, to the device characteristics on level Y is *under the control of the designer* through choice of device geometry as embodied in the mask set. Thus, to realize a resistance of a specific value the designer makes a suitable choice of its length and width. The designer also has some control over the correlation between the Y variables: for example, the location of two transistors in close proximity can introduce a (probably desirably) high correlation between their base-emitter voltages for identical quiescent conditions. Transformation to the highest level (Z) descriptive of circuit performance is governed by Kirchhoff's laws, and can be simulated by means of a circuit analysis package.

Unwanted variations occur at all levels, and lead to variations in the electrical performance of the integrated circuit. It is impossible, for example, to control a fundamental process variable (V) such as temperature with precision. Also, since the dimensions and registration of the masks within the mask set cannot be guaranteed except within limits, the transformation between levels X and Y further contributes to tolerances upon device parameters. Such tolerances, of course, lead to unwanted variations in circuit performance. It can be appreciated from Figure 6.4, therefore, that the tolerance behaviour of an integrated circuit is complex. To provide some appreciation of the complexity of the tolerance characteristics at level X, Figure 6.5 shows and explains a three-level statistical model relevant to the Monte Carlo analysis of an integrated circuit.

The figure relates to the statistical variations of a material electrical parameter (Level X of Figure 6.4) associated with an integrated circuit. Assume that we are concerned with sheet resistivity (RS); measurements of this parameter on all chips from the same wafer will, as shown in Figure 6.5(a), be characterized by a distribution representing the variation between *average* chip values. The distribution will be assumed to be Gaussian, with standard deviation σ_{bc} (bc = between chip). In a Monte Carlo analysis the parameter value

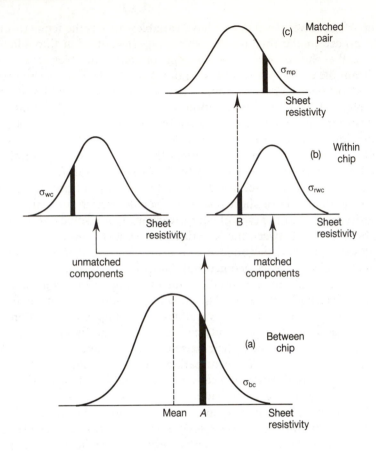

Figure 6.5
A three-level statistical model for the material electrical properties of an integrated circuit.

associated with a *given* chip will be selected from this distribution: assume its value (which is now the mean value over the selected chip) is equal to *A*. The variation of the parameter *within* the chip (i.e. across the components on the chip) is characterized (see Figure 6.5(b)) by a different Gaussian distribution of standard deviation σ_{wc} (wc = within chip). The value of sheet resistance for a *single component* would therefore be selected from this distribution. However, if ('twinned') components have been placed alongside each other for good matching, the within chip variation will be much reduced, as shown by the 'reduced within chip' distribution with standard deviation σ_{rwc}. Thus, in simulating the effect of sheet resistivity variations, a value (*B*) would be selected randomly from this (rwc) distribution. Finally, even for closely matched components there is (Figure 6.5(c)) *some* degree of mismatch between them, and the statistical nature of this mismatch is represented by a third-level distribution characterized by σ_{mp} (mp = matched pair). It must, of course, be borne in mind that this model refers only to level W of the model of Figure 6.4, so that unwanted variations in mask dimensions additionally determine the final variation of electrical parameter values at level Y.

6.3.1   Design freedoms

With respect to design freedoms, certain essential differences between discrete-component and integrated circuit design are immediately apparent. First, the designer exercises his design freedoms within the transformation between levels X and Y, rather than directly at the level (Y) of electrical device characteristics. Second, nominal parameter values can be chosen from an essentially continuous range: a required change in a resistance is easily effected by an appropriate change in its length or width or both. Third, tolerances can also be chosen anywhere between limits: a long and very thin resistor will exhibit a resistance far more sensitive to variations in linear dimensions than a square resistor would be. Fourth, strong correlations exist between parameter values associated with devices on the same chip: indeed, the success of a particular integrated circuit technique may depend fundamentally on the skilful exploitation of such correlations (for example, the correlation between capacitances in switched-capacitor circuits). Figure 6.6 illustrates the relation between the spaces of process parameters, electrical device parameters and circuit performance, and indicates the transformation within which the designer exercises his design freedom. Normally, the integrated circuit fabrication process cannot be tailored to suit a single mass-produced circuit, so that design at level V is not normally permitted.

With regard to the task of estimating the tolerance effects in integrated circuits it should be noted that, in contrast with the situation obtaining for discrete devices, more is usually known about the parameter distributions associated with integrated circuits, especially if design and fabrication are carried out within the same company. This is achieved by the constant monitoring of the integrated circuit fabrication process by measurements made on test chips associated with most or all wafers that pass through the process as described in Section 4.3.1. Knowledge concerning these distributions and their correlations can be incorporated in databases which are accessed by the programmes used to simulate circuit performance, since only in this way will a satisfactory estimate of the effect of tolerances be obtained (Rankin, 1982).

The *objectives* of the integrated circuit designer also differ in some respects from those of the designer of discrete circuits. The manufacturing yield of an integrated circuit is jointly determined by two factors, with each of which a 'partial yield' can be associated. The **process yield** is that partial yield which is degraded by material

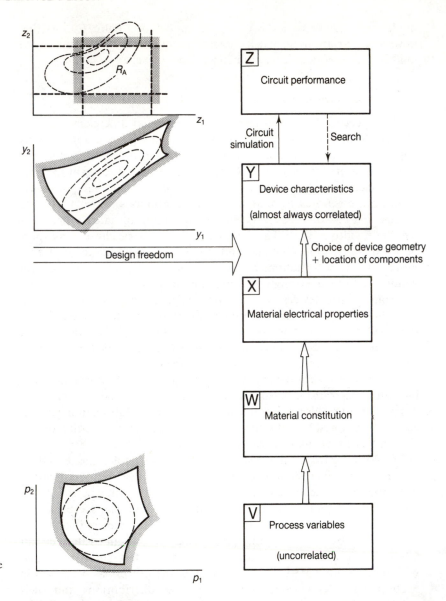

Figure 6.6
The design freedom available to the designer of an integrated circuit.

imperfections, catastrophic failures and processing defects, and is often the dominant factor in determining the manufacturing yield of an integrated circuit. The other factor, known as **parametric yield**, is what we have previously referred to simply as 'yield', and is determined purely by the unwanted (but not catastrophic) variations in electrical parameter values. Usually, in integrated circuit design, the objective is to maximize the parametric yield and, if at all possible, maintain it at 100%, but the satisfaction of performance specifications is not the only

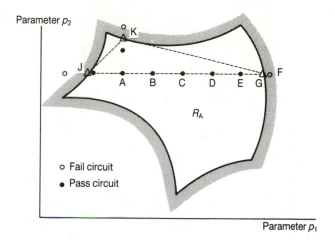

Figure 6.7

A search, involving only circuit analyses, to find the boundary of the region of acceptability. The triangle GJK is a first approximation to R_A.

objective of integrated circuit design. It is highly desirable, for example, to minimize the total chip area occupied by the design (since a reduced chip area enhances the process yield) and this objective may be in conflict with 100% parametric yield.

In view of the complexity of integrated circuit design it is not surprising that most attention has been paid to the tolerance design of discrete component circuits. However, a successful example of the application of tolerance design to an integrated circuit (Knauer and Pfleiderer, 1982) has already been discussed, and will be referred to again in Chapter 7. Another example of integrated circuit design (Jones and Spence, 1984) involved a memory cell whose area, power dissipation and delay time were required to be minimized (according to a weighting distribution) under the constraint of 100% parametric yield; an appreciation of tolerance design methods was a vital component in the method finally devised. A third example, again of an (NMOS) integrated circuit design, is provided by Hocevar *et al.* (1984). Recently, the application of tolerance design concepts to integrated circuits is increasingly being reported (see, for example, the January 1986 issue of *IEEE Transactions on the CAD of Integrated Systems*).

6.4 The deterministic approach

One method* of identifying the whereabouts of the region of acceptability, identified earlier as an essential task in tolerance design, is illustrated in Figure 6.7 and is called the simplicial approximation

---

* So that this chapter can be read without prior reference to earlier chapters, some brief repetition of concepts introduced earlier will occur.

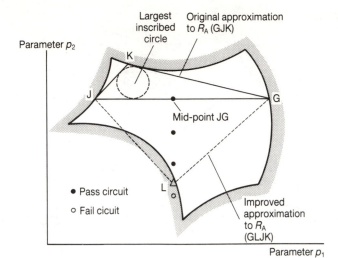

Figure 6.8

Expansion of the approximation to R_A to include a new point (L) identified by a search from the midpoint of the largest side of the original approximation (GJK) tangential to the largest inscribed circle.

method (Director and Hachtel, 1977). Suppose that a circuit represented by point A in parameter space has been designed, and that simulation by means of a suitable circuit analysis package has shown that it satisfies all performance specifications: it is therefore called a pass circuit. Being a pass circuit, it is known to lie inside R_A. A search for the boundary of R_A is now carried out by varying one parameter (p_1 in Figure 6.7) while maintaining all others fixed; at each step in the search (i.e. at points B, C, D, E and F) a circuit analysis is carried out to determine whether the circuit is a pass or fail circuit. Since the circuit passes at E but fails at F, interpolation of the performances whose boundary has been encountered establishes a point (G) which to a good approximation lies on the boundary of R_A. A similar search is now undertaken in the opposite direction to identify the point J as being on the boundary and (say) along the positive p_2 direction to identify point K. The polygon defined by the points G, J and K is taken as a first approximation to the region of acceptability.

As with the example shown in Figure 6.7, a triangular approximation of a two-dimensional region of acceptability is unlikely to be satisfactory. To effect an improvement the following algorithm may be used. First, the largest possible circle is inscribed within the polygon (Figure 6.8). Of the sides of the polygon it touches, the longest (JG) is identified, since this may well be the sector which is the poorest approximation to the boundary of R_A. From the mid-point of this side, and perpendicular to it in a direction away from the interior of the polygon, a new search direction is defined. A search, similar to the ones that identified points G, J and K, is now carried out to find another point (L) on the boundary of R_A. The original polygonal approximation to R_A is now expanded to include the new point L. This process can be repeated as many times as necessary to obtain the required approximation to R_A: we shall refer to this approximation as R'_A. It will be appreciated that, in Figures 6.7 and 6.8, a two-parameter example has been chosen for ease of illustration. In a circuit with many toleranced parameters the algorithm is essentially unchanged; now, for example, R'_A is a multidimensional polygon which, at each iteration, is expanded about the side of greatest area tangential to the largest inscribed hypersphere.

Under the assumption that the region of acceptability is convex (which is clearly not the case in Figure 6.8), the polygon R'_A lies entirely within R_A. A lower bound on the yield can then be obtained by determining the volume of R_T that lies inside R'_A, taking due account of the probability distribution of the parameters p_1 and p_2. One means of obtaining an estimate of the yield is to generate Monte Carlo sample points within R_T in accordance with the known probability distributions of p_1 and p_2, and classify them as pass or fail according to whether they lie inside or outside R'_A. No costly circuit analyses are involved, so a large number of points can be used to improve the accuracy of the estimate. It must be borne in mind, however, that R'_A itself is only an approximation to R_A.

A knowledge of R'_A can provide useful information regarding tolerance design. For example, if the axes p_1 and p_2 are suitably scaled, the joint probability density function contours will be circular, and the design centre corresponding to maximum yield will, to a good approximation, be the centre of the largest circle that can be inscribed within the scaled R'_A. The identification of this point is a standard task in linear programming. Also, the size of the largest inscribed circle can yield useful information concerning component tolerances.

While the deterministic approach, as described in outline above, appears simple in both concept and execution, it has several disadvantages:

(1)   By far the most serious disadvantage is that the method suffers from the curse of dimensionality in that, while the computational effort involved may be quite reasonable for the two-dimensional example shown in Figure 6.7, it escalates extremely rapidly as the number of adjustable toleranced parameters increases. Indeed, it is normally inadvisable for the method to be applied to circuits having more than five toleranced parameters.

(2)   Some aspects of the method assume the *convexity* of R_A, so that there is a danger of significant error if, as in Figure 6.8, parts of the R_A surface are concave. It is highly likely, though not proven except for special cases of resistive circuits, that closely separated upper and lower specifications give rise to complementary convex and concave surfaces of R_A in parameter space.

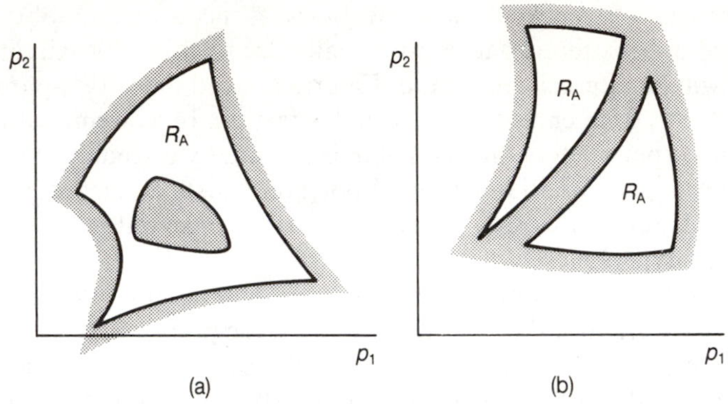

Figure 6.9
(a) A 'black hole' containing circuits failing the specifications.
(b) A disjoint region of acceptability.

(3)   The presence of 'black holes' within R_A (Ogrodski *et al.*, 1980), and disjoint acceptable regions (Figure 6.9), immensely complicates the determination of the boundaries of R_A.

Despite the fact that a great deal of research has been directed towards the improvement of deterministic methods of tolerance design, it appears to be generally agreed that, because of the curse of dimensionality, they are suited only to circuits of limited dimension.

6.5   Statistical exploration

It is the so-called statistical exploration approach to tolerance design with which this book is principally concerned, for reasons that will be outlined in this section. In the statistical exploration approach the actual circuit manufacturing process is simulated by making a random selection of component parameter values and then evaluating, by means of a circuit analysis package, the performance of each resulting circuit; the procedure is known as Monte Carlo analysis. The yield is then estimated from the number of circuits found to pass the specifications (Figure 6.10). As discussed in Chapter 4, an important property of such a Monte Carlo analysis is that the accuracy of the yield estimate $\hat{Y}$ is independent of the number of component parameters. However, the accuracy of the estimate *is* dependent on the number of samples used in the Monte Carlo analysis and only increases with the square root of the number of samples. For example, for an estimated yield of 60%, 100 circuit analyses will only provide 95% confidence on an estimate within plus and minus 10%: to find Y within 5% of the true answer, though still with only 95% confidence, would require 400 circuit analyses.

The information gained from a Monte Carlo analysis can be used to deduce some aspects of the whereabouts of R_A, at least

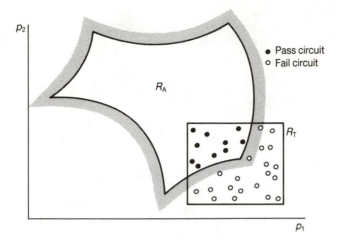

Figure 6.10
Statistical exploration involving Monte Carlo analysis: the simulation of randomly selected circuits within the tolerance region. The estimated yield is the fraction (here 10/30 or 33%) of simulated circuits which satisfy all the specifications.

sufficiently to suggest a possible improvement. Thus (Figure 6.11) it is reasonable to suppose that if the centres of gravity (CP and CF) of the pass and fail samples are located, the yield is likely to increase if the tolerance region is moved (i.e. the nominal parameter values are changed) along the line joining these two centres of gravity, and in a direction away from the centre of gravity of the fail circuits. Choice of a suitable displacement defines a new tolerance region whose yield is hopefully increased over that of the initial tolerance region (Figure 6.12). Clearly, many variations on the above theme are possible, and some will be described in Chapters 7, 8 and 9. For example, since one objective of the Monte Carlo analysis is to establish the relative positions of pass and fail samples, the samples could be selected within an inflated tolerance region that might better be termed a search region.

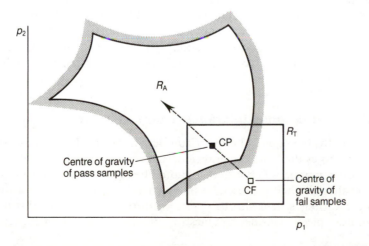

Figure 6.11
The dashed arrow joining the centres of gravity of the pass and fail samples is a potentially successful direction in which to move R_T if the yield is to be increased.

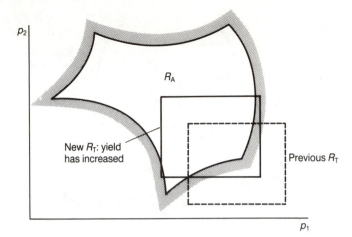

Figure 6.12
Movement of the tolerance region (i.e. the choice of new nominal values for p_1 and p_2) along the direction identified in Figure 6.11 may lead to an increase in yield.

6.6   Comparison of approaches

In the same way that the outcome of a Monte Carlo analysis is independent of the dimensionality of the problem (i.e. the number of toleranced parameters), all the evidence available points to the fact that tolerance design methods based on statistical exploration are also, in a computational sense, essentially independent of the dimensionality of the problem. In other words, the number of circuit analyses required to design centre a circuit is liable to be no more for a circuit of 50 components than it is for a simple potential divider. This remarkable property of statistical exploration methods, when combined with their ability to handle robustly the variety of features (e.g. concavity, black holes) that can be associated with the regions of acceptability of practical circuits, has been responsible for their widespread use in industry. Table 6.1 summarizes the advantages and disadvantages of the deterministic and statistical approaches to tolerance design, though the implied 'either/or' nature of the choice should be tempered with the remark that there may well be situations in which a blend of the two methods may turn out to be advantageous (Tahim and Spence, 1979).

6.7   Evaluation of tolerance design schemes

Eventually, the potential user (circuit designer or CAD specialist) must be able to evaluate any tolerance design scheme that is offered. Table 6.2 lists a number of criteria that should be borne in mind during such an evaluation, and can usefully be taken as a guide for the assessment of the various schemes to be described in the remainder of this book. For example, in considering Question 1(a) it is confirmed by experience

Table 6.1 The advantages and disadvantages of the deterministic and statistical exploration approaches.

Deterministic	Statistical exploration
Advantages	*Advantages*
Good for 'worst case'	Independent of number of
Efficient for small	toleranced parameters
number of parameters	No assumptions about the
Mathematical rigour	nature of R_T, R_A
	Yield accuracy predictable
Disadvantages	*Disadvantages*
Only useful for small	Expensive to guarantee
number (up to 5–8)	high yields (e.g. 99%)
toleranced parameters	Largely heuristic
Maximizes only a lower	algorithms employed
bound on yield	
Limiting assumptions about	
nature of R_T, R_A and	
probability distributions	
No means of predicting	
the accuracy of the yield	
estimate: only a lower bound	
can be found under certain	
conditions (e.g. if R_A is convex)	

Table 6.2 Some criteria to assist the evaluation of a tolerance design scheme.

(1)   How general is the method?
    (a)   Are assumptions made about R_T, R_A and $\phi(p)$?
    (b)   What is the maximum number of toleranced parameters?
    (c)   What is the maximum number of design variables?

(2)   What information does it give me?
    (a)   Will it do design centring, tolerance assignment, ... and for what
        aspects of circuit performance?
    (b)   Can I easily change the specifications?
    (c)   Is it better for worst-case design (e.g. 100%) or less than 100% yield?
    (d)   How accurate is the method?
    (e)   What are its convergence properties like?

(3)   What does it cost to use?
    (a)   How many circuit analyses to find the yield or R_A?
    (b)   How many circuit analyses (typically) to improve the yield?
    (c)   How severe are the computational overheads?

(4)   Ease of implementation
    (a)   Are the mathematics comprehensible?
    (b)   Can the method use an already existing circuit analysis package?

that the statistical exploration approach is far superior to the deterministic approach. Companies wishing to make reasonably immediate use of tolerance design will be concerned with the answer to Question 4(b): if an existing circuit analysis package can be used without modification, and only a small amount of 'bolt-on' software need be written, then the ease of implementation and the reliability of the results will be considerably enhanced.

6.8   Further reading

The following four articles in the field of tolerance design may be useful to those requiring a more detailed overview. All four articles include extensive and useful bibliographies for those wishing to dig deeper. The first three are reviews while the fourth presents the results of a comparison of different published methods applied to a set of common circuit problems.

Brayton, R. K., Hachtel, G. D. and Sangiovanni-Vincentelli, A. L. (1981) 'A survey of optimization techniques for integrated circuit design.' *Proc. IEEE*, **69**(10), 1334–1363 [Good overview of both approaches, discussion of interactive systems for optimization, and useful list of references.]

Director, S. W. and Vidigal, L. M. (1981) 'Statistical circuit design: a somewhat biased survey.' *Proc. Eur. Conf. Cct. Theory Design (ECCTD)*, 1981, The Hague, pp. 15–24 [Discusses both deterministic and statistical methods, with emphasis on the former.]

Spence, R., Gefferth, L., Ilumoka, A. I., Maratos, N. and Soin, R. S. (1980) 'The statistical exploration approach to tolerance design.' *Proc. IEEE Int. Conf. Ccts. Computers*, Oct. 1980, New York, pp. 582–585 [Surveys statistical exploration techniques.]

Wehrhahn, E. and Spence, R. (1984) 'The performance of some design centering methods.' *Proc. IEEE Int. Sym. Ccts. Sys.*, Montreal 1984, pp. 1424–1438.

CHAPTER 7

Simple Methods using Performance Calculations

OBJECTIVES

If tolerance analysis predicts that the manufacturing yield of a mass-produced circuit will be substantially less than 100%, a common requirement is that redesign should take place in an attempt to maximize the yield. In some cases it will be possible, by making changes to the nominal values of the components (while leaving their tolerances fixed), to achieve 100% yield, while in others the specifications will turn out to be so tight that the desirable objective of 100% yield cannot be achieved. In this chapter two yield maximization algorithms are presented, one for each of the two situations. The methods differ from others described in Chapters 8 and 9 in their sole dependence upon computed values of circuit performance. The algorithms are characterized by the fact that, apart from some relatively trivial calculations, most of the computational effort can be carried out by the sort of electronic circuit analysis package that many laboratories already possess. For this reason the implementation of either of the algorithms in a practical industrial environment is quite straightforward. Both the algorithms have been subjected to a great deal of practical testing, and have been found to be robust in operation and reasonably economic to apply.

7.1   The problem: yield maximization

The problem, common to the two algorithms to be described, is that of maximizing the manufacturing yield of a mass-produced circuit. The maximization is to be achieved by adjusting the nominal values of the component parameters, the parameter tolerances remaining fixed. Such a constraint upon tolerances is, as we have remarked earlier, entirely realistic. For example, in hybrid and integrated circuits, the parameter tolerances are determined by the inevitable and essentially unalterable variations in the fabrication process. Even with discrete circuits, for which parameter tolerances can freely be chosen, design centring is normally carried out prior to tolerance assignment under the reasonable assumption that the design centre leading to maximum yield is unlikely to be strongly dependent upon parameter tolerances, and it is preferable to expand or contract tolerances only after the design is centred.

The problem we are discussing in this chapter can be stated formally as

$$\begin{aligned}
&\text{Maximize } Y(\phi(p, p^0)) \\
&\quad \text{by choice of } P^0 = p_1^0, p_2^0, \ldots, p_K^0 \\
&\quad \text{for fixed tolerances } t = t_1, t_2, \ldots, t_K
\end{aligned} \tag{7.1}$$

As before, ϕ is the K-dimensional component parameter probability density function, with p^0 representing the mean (nominal) values. One possible example of ϕ is the truncated multidimensional Gaussian, with the truncation carried out at the 3σ points of the distribution. In that case $t_i = 3\sigma_i$, where t_i is the tolerance on the ith parameter and σ_i is the standard deviation associated with the parameter. Another example of ϕ, and one to which we shall refer in this chapter, is the multi-

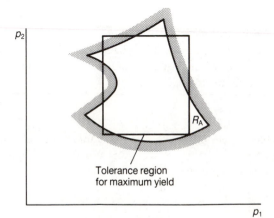

Figure 7.1

The circuit and specifications are such that 100% yield cannot be obtained with the existing tolerances.

Tolerance region
for maximum yield

R_A

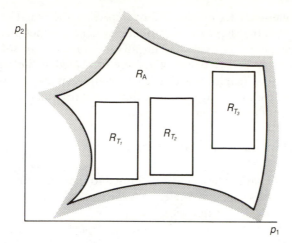

Figure 7.2
Three different designs leading to 100% yield.

dimensional uniform distribution in which, for each parameter, all values between the tolerance limits are equally likely to occur.

As it is stated, Objective 7.1 is an unconstrained optimization. It is true that, in practice, there will be constraints on the designable parameters: for example, there will be limits to the resistance values that can be achieved with a diffused resistor. However, the yield maximization exercise normally starts from a feasible nominal design and such constraints are seldom encountered. There is therefore little profit to be had from complicating Objective 7.1 any further.

For a given circuit with specific parameter tolerances it may turn out not to be possible to achieve 100% yield. Such is the case illustrated in Figure 7.1. In this situation it is reasonable to refer to the yield maximization as design centring, since the maximum yield will be obtained when the nominal design lies at the (normally unique) 'centre' of the region of acceptability. For another circuit it may be possible not only to achieve 100% yield, but to do so (Figure 7.2) with a number of alternative nominal designs. In other words, the optimum design is not unique. It seems inappropriate in this situation to speak of 'centring'. Instead we refer to the exercise of choosing the nominal to achieve 100% yield as worst-case design. The term worst-case design is used because, for 100% yield, even the circuit with the worst values of performance must meet all the specifications. In worst-case design the objective may be stated formally as

$$\text{Find } P^0 = P^* \text{ such that } Y(\phi(P, P^*)) = 1 \qquad \textbf{(7.2)}$$

It is common to encounter the worst-case design problem with discrete component circuits for which 100% yield can normally be achieved if sufficiently small tolerances are selected.

In one sense Expression 7.2 overspecifies the worst-case design problem, since the shape of the probability density function ϕ is of no consequence. For any two different pdfs the same design solution will suffice provided the two distributions are truncated at the same limits. It may be more helpful, instead, to formally express the worst-case design objective as merely requiring that the entire tolerance region lie anywhere within the region of acceptability:

$$\text{Find } P^0 = P^* \text{ such that } R_T(P^*) \subset R_A \tag{7.3}$$

If the nominal parameter values are limited to those in a discrete set (for example the E24 series of 'preferred values') then a constraint to this effect can be added to Expression 7.3.

Another consideration of moderate importance is that it may be the tolerances *relative* to the nominals which are fixed and not their *absolute* values. For instance, the absolute value of a parameter whose tolerance is fixed at 10% will depend on the actual nominal value of that parameter. Therefore, when designs with different nominals (design centres) are considered, the absolute values of the tolerances and hence the size of R_T also changes. In some situations, for example with integrated circuits, the absolute value of the tolerance remains fixed: an example is the over/under-etch ΔR in the fabrication of integrated circuit resistors. In the subsequent discussions in this chapter we may have to distinguish whether absolute or relative tolerances are being considered as fixed.

It is true that all the objectives 7.1 to 7.3 set out above have the maximization of yield as their goal. In practice, however, it has been found that, according to whether 100% yield can or cannot be achieved, different classes of algorithm may be more effective. Thus, we shall speak below of design centring when 100% yield cannot be achieved, and worst-case design when it can. Since the designer doesn't usually know at the outset whether or not 100% yield is possible, some careful transfer between algorithms may be necessary.

7.2 A conventional algorithm

One straightforward approach to the maximization of yield is illustrated in Figure 7.3 for the case of a circuit with two designable component parameters. As a first step yield is estimated for three points A, B and C defining an equilateral triangle in parameter space. For convenience we refer to this triangle as the current triangle. Now let us say for instance that the estimated yield $\hat{Y}_B$ was found to be less than both $\hat{Y}_A$ and $\hat{Y}_C$.

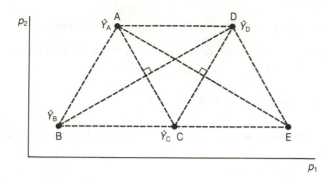

Figure 7.3
The simplex method of
optimization.

That is, B is the worst vertex in the current triangle. In the next step, B (the worst vertex) is discarded for another exploratory point D, obtained by reflecting B in the opposite face AC of the current triangle.

ACD now forms the current triangle for further exploration. Yield is evaluated for the new point D. If for instance point A were now to become the worst vertex of the current triangle (that is, $\hat{Y}_A$ was less than both $\hat{Y}_C$ and $\hat{Y}_D$), a new exploratory point E would be obtained by reflecting A in the face CD.

This procedure would continue until the new exploratory point was also the worst vertex. For example if $\hat{Y}_E$ was found to be less than $\hat{Y}_C$ and $\hat{Y}_D$ then, according to our previous procedure, the new exploratory point should be A. But we have just discarded A. One way of proceeding further is to reduce the size of the triangle, find the reflection of E in the opposite side of this new exploratory triangle and continue. There are several variants of this strategy, which is basically the simplex method of optimization. The term 'simplex' arises from the fact that a polygon in higher dimensions is referred to as a simplex, and indeed a triangle is a two-dimensional simplex. The method belongs to a family of optimization techniques referred to as direct search methods; they are to be distinguished from gradient-based methods which, unlike the former, employ the gradients of the objective function.

We observe that, in three-parameter space, the simplex would be a pyramid, and four values of yield would have to be considered at each step of the algorithm. More disturbingly, for a realistic circuit of 50 parameters a total of 51 estimates of yield would have to be obtained before even the first exploratory movement in parameter space could be made. Clearly, the simplex method appears most unattractive for yield maximization since the estimate of yield for each point in parameter space can involve at least 100 circuit analyses. Nevertheless, the literature provides examples of such an approach (Becker and Jensen, 1977) using a kindred method, pattern search (Hooke and Jeeves, 1961).

7.3   Design centring by statistical exploration

One successful method of design centring is based on the fact that a Monte Carlo analysis performs two useful functions:

(1)   the estimation of manufacturing yield;

(2)   by virtue of its pseudo-random sampling of parameter space, the provision of spatial information about the region of acceptability.

The first function was adequately discussed in Chapter 4. The second, prior to a later detailed discussion, can easily be outlined by pointing to the fact that, for each of the Monte Carlo sample points which correspond to locations in parameter space, we know whether the corresponding circuit was acceptable or not. By taking a bird's-eye view of this information we can identify potentially profitable directions in which to move the tolerance region. In describing the first algorithm, called the **centres of gravity algorithm** for design centring, we shall distinguish between these two functions of a Monte Carlo analysis.

The centres of gravity (CoG) method can be illustrated by reference to Figure 7.4. In (a) we see the tolerance region associated with the initial circuit design so positioned with respect to the region of acceptability that there is potential for yield increase. The overlap $(R_T \cap R_A)$ between R_T and R_A corresponds to acceptable circuits, and the remainder of R_T (that is, $R_T \cap \bar{R}_A$) to failed circuits. Now assume that a Monte Carlo analysis is carried out, as shown in (b). One outcome is an estimate $\hat{Y}$ of the yield; the other is the location of the pass and fail samples, some of which are indicated in the figure. Now assume that the centres of gravity of the pass and fail circuits (called CP and CF respectively) are calculated from the results of the Monte Carlo analysis (Figure 7.4(c)). With an infinite number of samples these centres of gravity will be identical to the actual centres of gravity of the two regions when weighted according to the applicable component probability density functions. It can also be shown that the two true centres of gravity and the nominal point will be colinear whatever the component pdfs (Antreich and Koblitz, 1980).

Intuition suggests that the yield will be increased if the nominal design (p^0) is moved along the line joining CP and CF, in the direction away from CF, as shown in Figure 7.4(d). This surmise is, in fact, the basis of the centres of gravity design centring method. The extent of the required movement is certainly not obvious, since the statistical exploration carried out by the Monte Carlo analysis has occurred entirely within R_T, and we have no information as to what the yield will be for a new R_T. Also, we know that we cannot simply search along this direction until a maximum yield is found, because this would require a

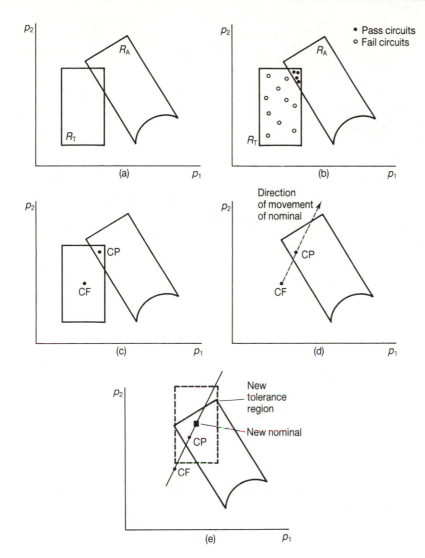

Figure 7.4
The centres of gravity method of design centring.
(a) The initial tolerance region.
(b) Monte Carlo sampling within R_T.   (c) Identification of the centres of gravity of pass and fail samples.   (d) Direction in which nominal (centre of tolerance region) can be moved to increase yield.
(e) Selection of new design.

new Monte Carlo analysis at each of several points along that direction. Rather, a single step is taken: the nominal design is repositioned at that location for which the yield might be expected to be usefully enhanced (Figure 7.4(e)). The procedure is then repeated, starting with a fresh Monte Carlo analysis (initially to see if the yield has increased) and then as many times as the designer might think fit. Naturally, if the Monte Carlo analysis carried out for the new design indicates a decrease in yield, the procedure may be terminated. Thus, the CoG algorithm is iterative, with the essential structure shown in Figure 7.5(a). Figure 7.5(b) shows the result of applying the CoG method to a

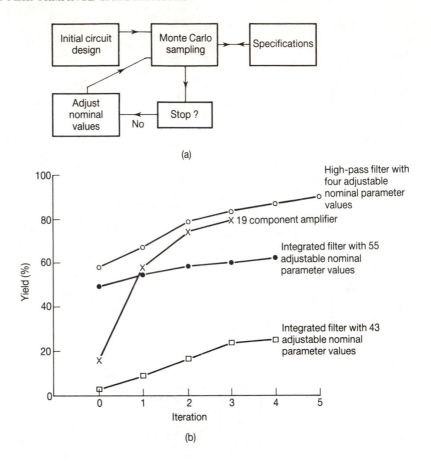

(a)

(b)

Figure 7.5
(a) Essential structure of the centre of gravity design centring algorithm.   (b) The result of applying the centres of gravity design centring algorithm to a selection of circuits.

selection of circuits, and indicates that useful increases can be obtained in few iterations.

We now discuss the many factors that must be taken into consideration when applying this apparently simple algorithm.

7.3.1   Step length

There is no consensus of opinion about the distance that the nominal design should be moved along the direction to become the new trial design. In the earliest investigations by Soin and Spence (1980) the distance moved was identical to the separation of the centres of gravity. Recently, some workers have used the centre of gravity (CP) of the pass circuits as the new design, while Ibottson *et al.* (1984) have obtained good results by moving the nominal a fraction $(1 - \hat{Y})$ of the distance between the centres of gravity. A similar suggestion was made by Soin and Spence (1980). Formally, if the new design can be expressed as

$$P^0_{new} = P^0_{old} + \lambda(CP - CF)$$

then these alternative choices of step length can be expressed as

(1) $\lambda = 1$

(2) $P^0_{new} = CP$

(3) $\lambda = (1 - \hat{Y})$

It is entirely possible that new and successful guidance regarding the choice of step length will continue to emerge.

7.3.2   Correct ranking of yield estimates

Let us consider two iterations of the CoG method, and assume that the results are as shown in Figure 7.6(a). Apparently, the yield has shown a substantial increase due to the choice of a new set of nominal parameter values, and so it is with confidence that we might accept this new design as a distinct improvement over the old one. That this confidence might be misplaced is illustrated in Figure 7.6(b). The first Monte Carlo analysis estimates a yield of $\hat{Y}_1$, but we know from Chapter 4 that all we can say is that, within a confidence level of (say) 95%, the actual yield will lie between two limits that can easily be computed, and which are indicated in the figure. Thus the *actual* yield Y_1 for the first design could, as in the figure, lie well above the estimate (or even *outside* the confidence limits). Similar considerations apply to the second Monte Carlo analysis; here, as Figure 7.6(b) shows, the actual yield Y_2 may be well below its estimated value $\hat{Y}_2$. Thus, for the case shown, in contrast

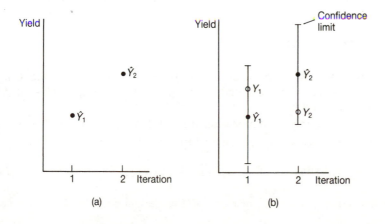

(a)   (b)

Figure 7.6
Showing that an increase in estimated yield can conceal a decrease in actual yield.

to an apparent increase in yield, the new nominal design has led to a decrease.

What is needed, therefore, is some means of computing a measure of confidence that an estimated change $\widehat{\Delta Y}$ ($= \hat{Y}_2 - \hat{Y}_1$) is of the same sign as the actual change $\Delta Y = (Y_2 - Y_1)$, even though it may be numerically different. In other words, we seek some measure of the confidence in the correct ranking of yield estimates. Indeed, it should be remarked that in the early stages of design centring by the CoG method it is the *change in yield*, rather than the yield itself, which is of primary concern.

7.4 Ranking of yield estimates

To obtain some insight into the problem, assume that Monte Carlo analyses have been carried out within two tolerance regions R_{T_1} and R_{T_2} in parameter space (Figure 7.7). We denote the estimated yields by $\hat{Y}_1$ and $\hat{Y}_2$ respectively, and we shall assume that the estimate $\widehat{\Delta Y}$ of the change ΔY in yield ($\Delta Y = Y_2 - Y_1$) is obtained simply by subtracting the two yield estimates:

$$\widehat{\Delta Y} = \hat{Y}_2 - \hat{Y}_1$$

Like $\hat{Y}_1$ and $\hat{Y}_2$, the quantity $\widehat{\Delta Y}$ is also a random variable, in the sense that if the experiment involving the two Monte Carlo analyses were to be repeated a number of times, different values would most likely be obtained for $\widehat{\Delta Y}$. Since $\hat{Y}_1$ and $\hat{Y}_2$ have Gaussian distributions then $\widehat{\Delta Y}$, the difference between them, will also have a Gaussian distribution, as depicted in Figure 7.8(a).

To characterize the Gaussian distribution of $\widehat{\Delta Y}$ shown in Figure 7.8(a) we need to establish its mean and variance. The mean of $\widehat{\Delta Y}$ is the difference between the means of $\hat{Y}_1$ and $\hat{Y}_2$; but since these means are the true values of Y_1 and Y_2 respectively, we conclude that

$$E(\widehat{\Delta Y}) = \Delta Y$$

In other words, if the experiment involving the two Monte Carlo analyses were to be repeated many times, the average value of $\widehat{\Delta Y}$ would tend to ΔY.

Employing another standard result from probability theory, the variance of the probability density function of $\widehat{\Delta Y}$ will be given by

$$\mathrm{var}(\widehat{\Delta Y}) = \mathrm{var}(\hat{Y}_1) + \mathrm{var}(\hat{Y}_2) - 2\,\mathrm{cov}(\hat{Y}_1, \hat{Y}_2) \tag{7.4}$$

where $\mathrm{cov}(\hat{Y}_1, \hat{Y}_2)$ is a measure of the dependence between the two

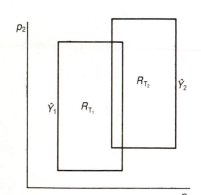

Figure 7.7
Monte Carlo analyses performed for two tolerance regions.

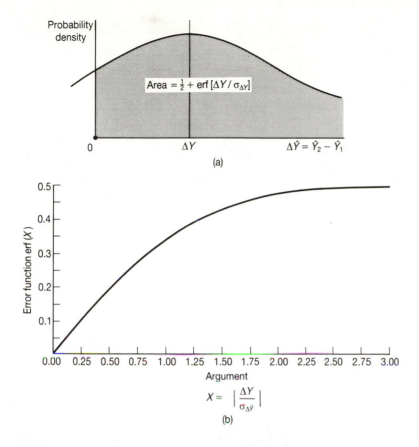

Figure 7.8
(a) Probability density distribution of the difference in the yield estimates of Figure 7.7.   (b) Plot of the error function erf(x).

estimates $\hat{Y}_1$ and $\hat{Y}_2$. We shall return shortly to consider this expression in detail.

Figure 7.8(a) shows the probability density function of the random variable $\hat{Y}$ for the case where $\widehat{\Delta Y}$ is positive. The confidence of correctly ranking $\hat{Y}_2$ and $\hat{Y}_1$ in this case is the probability that an estimate of ΔY will be positive. This probability is the area under that part of the curve between $\widehat{\Delta Y} = 0$ and $\widehat{\Delta Y} = \infty$. Fortunately, for Gaussian distributions, such an area has been expressed in terms of its mean and variance, and tabulated as the error function erf(.) (Larson, 1969). A plot of the error function is shown in Figure 7.8(b). The required area equals

$$1/2 + \mathrm{erf}\left[\frac{\Delta Y}{\sqrt{\mathrm{var}\,(\widehat{\Delta Y})}}\right]$$

Strictly, the computation of the area under the curve requires knowledge of both ΔY and $\mathrm{var}\,(\widehat{\Delta Y})$, the very quantities we do not know. The practical solution to this difficulty is the same as that adopted for

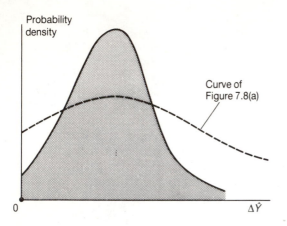

Figure 7.9
Confidence of correct ranking is increased by reducing $\sigma_{\widehat{\Delta Y}}$.

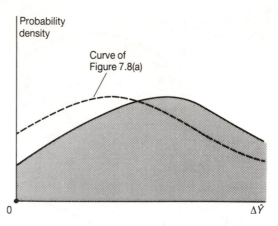

Figure 7.10
Confidence of correct ranking is increased by increasing the mean of $\widehat{\Delta Y}$.

calculating the confidence intervals for yield estimates in Chapter 4, namely, to use *estimates* of ΔY and $\mathrm{var}(\Delta Y)$. The former is simply the difference between the two yield estimates. The latter, if we assume initially that the covariance is zero, can be found from Equation 7.4:

$$\mathrm{var}(\widehat{\Delta Y}) = \mathrm{var}(\hat{Y}_1) + \mathrm{var}(\hat{Y}_2)$$

where $\mathrm{var}(\hat{Y}_1) = \hat{Y}_1(1 - \hat{Y}_1)/N$ and $\mathrm{var}(\hat{Y}_2) = \hat{Y}_2(1 - \hat{Y}_2)/N$, N being the number of Monte Carlo samples.

We have seen above how the confidence of correct ranking of two yield estimates can be computed. But how can it be improved? Two approaches are available. Since the error function is a monotonically increasing function of its argument, which in this case is the quotient $\widehat{\Delta Y}/\sqrt{\mathrm{var}(\widehat{\Delta Y})}$, the confidence of correct ranking will be enhanced either if $\widehat{\Delta Y}$ is larger or if the standard deviation $\sqrt{\mathrm{var}(\widehat{\Delta Y})}$ is smaller. These two situations are illustrated in Figures 7.9 and 7.10. Two sampling schemes called 'correlated sampling' and the 'common points scheme' aim to achieve these two effects respectively and are described in the following sections.

7.4.1   Correlated sampling

Correlated sampling is a means of introducing a high positive correlation between successive yield estimates, and is illustrated in Figure 7.11 which shows two tolerance regions for which the yields are to be

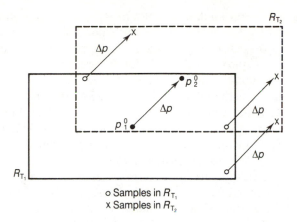

○ Samples in R_{T_1}
✕ Samples in R_{T_2}

Figure 7.11
Illustrative of correlated sampling.

estimated. First, a conventional Monte Carlo analysis is carried out within R_{T_1}. The pseudo-random points are selected from a distribution $\phi(P_1^0)$ centred on P_1^0 which is the centre of R_{T_1}. In this way an estimate of the yield Y_1 is obtained. We now turn to tolerance region R_{T_2}. One way of estimating the associated yield Y_2 would, of course, be to carry out a fresh Monte Carlo analysis using an entirely new set of random numbers to define the samples chosen from the component distribution $\phi(P_2^0)$. But an alternative and more desirable approach, and one which is based on the principle of correlated sampling, is to use the same pattern of samples in R_{T_2} as was used in R_{T_1}. For the situation being considered here, where R_{T_2} is exactly the same size as R_{T_1} but with a different centre point, correlated sampling involves taking every sample point in R_{T_1} and moving it by a distance Δp which is the displacement of the new design P_2^0 from the previous design P_1^0, as illustrated in Figure 7.11. Note that, with this approach, the sample points within R_{T_2} can be assumed to be just as satisfactory a sampling according to $\phi(p_2^0)$ as was the sampling within R_{T_1} from $\phi(p_1^0)$.

Correlated sampling, also called the method of common random numbers, is a general technique often applied where the results of Monte Carlo experiments on two similar situations are to be compared. The conventional method of effecting the scheme is to note the seed of the pseudo-random number generator (discussed in Section 4.3.2) at the start of the first Monte Carlo analysis. Then, before the start of the second Monte Carlo analysis, the random seed is deliberately set to the same value as at the start of the first Monte Carlo analysis.

The two yield estimates for R_{T_1} and R_{T_2} are obtained in the usual way, by counting the number of passes and dividing by the total number of samples. Correlated sampling introduces a useful relation between the two estimates. The consequences of using an identical sampling pattern are that, provided the tolerance regions are close (ΔP^0 is small), there is a positive correlation between the outcomes (i.e.

pass or fail) of corresponding samples in R_{T_1} and R_{T_2}. In other words, if one point in R_{T_1} is a pass, then the corresponding point in R_{T_2} is also likely to be a pass, since the points are in close proximity in component space. As a consequence, if the yield for R_{T_1} were to be overestimated, then the yield for R_{T_2} is also likely to be overestimated. This is fortunate, since even though the individual yields might not be estimated very accurately, the *difference* between them would be known quite accurately. In other words, the confidence of correct ranking of the two yields is increased over the level that would be obtained through two entirely independent Monte Carlo analyses.

Mathematically, this situation is represented by a positive value of $\mathrm{cov}(\hat{Y}_1, \hat{Y}_2)$ in Equation 7.4 and, consequently, a reduction in $\mathrm{var}(\widehat{\Delta Y})$, the 'uncertainty' in the value of the difference between the two yields. Such a reduction in $\mathrm{var}(\widehat{\Delta Y})$ is associated with a $\widehat{\Delta Y}$ distribution as shown in Figure 7.9, and consequently with an increased confidence in correct ranking. If indeed the Monte Carlo samples in the two tolerance regions were independent then the term $\mathrm{cov}(\hat{Y}_2, \hat{Y}_1)$ in Equation 7.4 would have been zero. A quantitative measure of this confidence is the area under the curve between $\widehat{\Delta Y} = 0$ and $\widehat{\Delta Y} = \mathrm{infinity}$. How this area can be computed in a specific case is considered in Section 7.8.

7.4.2   Common points scheme

When the common points scheme was first devised, its original objective was to achieve considerable savings in the computational effort associated with design centring by the centres of gravity method. However, the scheme can also lead to an increased confidence in the correct ranking of yield estimates.

The principle will be illustrated by the numerical example shown in Figure 7.12. First, we assume that the yield associated with tolerance region R_{T_1} is estimated by means of a conventional 100-sample Monte Carlo analysis, and found to be 78%. Now consider tolerance region R_{T_2} (which may represent a new design suggested by the CoG design centring method) for which the yield must also be estimated. Conventionally we would employ another 100 circuit analyses to achieve this estimate by Monte Carlo analysis. But we note that R_{T_1} and R_{T_2} overlap considerably: why not, therefore, re-use the samples in this overlap region? If we do so, noting that the overlap region contains 70 of the original 100 samples, then only 30 additional samples need be randomly selected within the region marked B in Figure 7.12 to make a total of 100 samples in the overall region R_{T_2}, and then analysed. In the numerical example shown we assume that, within region B, 25 passes and five fails are recorded. An estimate of the yield associated with R_{T_2}

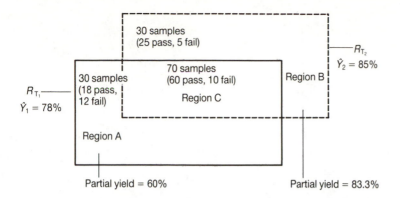

Figure 7.12

Illustrative of the common points scheme.

can now be obtained by combining the samples within **B** with those in the common points region **C**, so called because the samples associated with these points are common to the two tolerance regions. Using the numerical example of Figure 7.12 an estimate of 85% is obtained for the yield of R_{T_2}.

We see from this example that one advantage of the common points scheme is that the number of additional circuit analyses required to estimate Y_2 has been reduced from 100 to 30, a very considerable saving indeed. The saving in any particular situation, of course, is dictated entirely by the degree of overlap of the two tolerance regions. In a typical design centring exercise, 100 analyses might be associated with the first iteration, and then many less for each successive iteration. Figure 7.13 shows an example involving four iterations in which a total of 217 samples were employed in place of the 400 that would have been necessary if the common points scheme had not been used.

To examine the implication of the common points scheme for the confidence of correct ranking of yield estimates, it is necessary first to discuss the *partial yields* associated with the overlap (C) and non-overlap (A, B) regions identified in Figure 7.12. For the numerical example selected for illustration, estimates of these partial yields are

$$Y_A = 0.6 \qquad Y_B = 0.833 \qquad Y_C = 0.857$$

Since the samples in region C are common to the two tolerance regions, the difference $\Delta Y = (Y_2 - Y_1)$ in the yields of these tolerance regions is due entirely to the difference $\Delta Y_P = (Y_B - Y_A)$ in the partial yields associated with the non-overlap regions A and B. The difference ΔY_P in the partial yields is normally much greater than the difference ΔY in the true yields. For the particular example of Figure 7.12, $\widehat{\Delta Y_P} = 0.233$ and $\widehat{\Delta Y} = 0.07$. The general proof of this result is fairly straightforward. Denoting the volumes of regions A, B and C by V with an appropriate

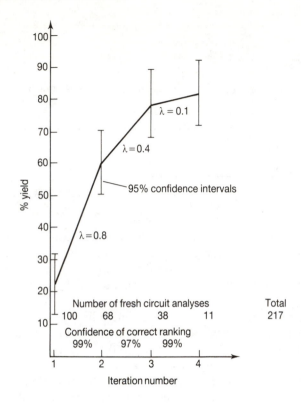

Figure 7.13
A typical yield trajectory
demonstrating the advantages of the
common points scheme.

subscript, we can express the yields Y_1 and Y_2 by

$$Y_1 = Y_A(V_A/V_T) + Y_C(V_C/V_T)$$

$$Y_2 = Y_B(V_B/V_T) + Y_C(V_C/V_T)$$

where V_T is the volume of each tolerance region. Thus, by subtraction, we obtain

$$\Delta Y = Y_2 - Y_1 = (V_B/V_T)Y_B - (V_A/V_T)Y_A$$

Recalling that $V_A = V_B$,

$$\Delta Y = (V_A/V_T)\Delta Y_P$$

from which it follows that, since $V_A/V_T < 1$, ΔY_P must always be greater than ΔY, by a factor V_T/V_A which, for the numerical example of Figure 7.12, is $100/30 = 3.33$.

We now recall, especially from Figures 7.8 and 7.10, that one way of enhancing the confidence of correct ranking is to increase ΔY; this we have evidently done by employing the common points scheme and directing our attention, not to $Y_2 - Y_1$, but to the difference ΔY_P in the partial yields of the non-overlap regions. But an increased yield is not, by itself, sufficient to enhance the confidence of correct ranking. As we saw earlier, this confidence is equal to an area under a distribution, as expressed by the error function. We must, therefore, additionally examine the variance of $\widehat{\Delta Y_P}$ and see how it is affected by the common points scheme. Again, to give some feeling for actual values, we shall use the numerical example of Figure 7.12. Since the samplings in A and B are uncorrelated,

$$
\begin{aligned}
\mathrm{var}\,(\widehat{\Delta Y_P}) &= \mathrm{var}\,(\hat{Y}_A) + \mathrm{var}\,(\hat{Y}_B) \\
&= Y_A(1 - Y_A)/N_P + Y_B(1 - Y_B)/N_P
\end{aligned}
$$

where, for the non-overlap regions, $N_P = 30$. Using estimates of yields in place of actual yields, we obtain

$$
\mathrm{var}\,(\Delta Y_P) = 0.008 + 0.004637 = 0.01263
$$

So

$$
|\widehat{\Delta Y_P}|/\sqrt{\mathrm{var}\,(\widehat{\Delta Y_P})} = 2.0726
$$

For purposes of comparison we now determine $\mathrm{var}\,(\widehat{\Delta Y})$ from

$$
\widehat{\mathrm{var}}\,(\widehat{\Delta Y}) = \hat{Y}_2(1 - \hat{Y}_2)/N + \hat{Y}_1(1 - \hat{Y}_1)/N
$$

With $N = 100$, and again using estimates of yields in place of actual yields, we find

$$
\mathrm{var}\,(\widehat{\Delta Y}) = 0.001716 + 0.001275 = 0.002991
$$

So

$$
|\widehat{\Delta Y}|/\sqrt{\mathrm{var}\,(\widehat{\Delta Y})} = 1.279
$$

Thus, for the present example, the ratio $\widehat{\Delta Y}/\sqrt{\mathrm{var}\,(\widehat{\Delta Y})}$ has been increased from 1.279 to 2.0726 by means of the common points scheme. When substituted into the expression for confidence of correct ranking introduced earlier, this represents an increase in the confidence of the correct ranking of regions R_{T_1} and R_{T_2} from 89.9% to 98.1%. The original and straightforward common points scheme described above was confined to uniform component pdfs. Recently, Stein (1986) has extended this scheme to more general component pdfs.

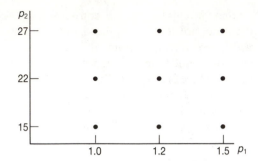

Figure 7.14
Possible nominal designs if parameters are restricted to 'preferred values'.

7.5   Worst-case design

As we know, worst-case design is the selection of a nominal design (and there may be more than one) such that all manufactured samples pass the specifications. In other words, 100% yield is the objective. This section describes one successful method – the cut method – of worst-case design (Wehrhahn, 1984). In presenting the method we shall assume, purely for illustration and without any restriction, that nominal component values can only take on 'preferred' values (Figure 7.14).

The essence of the cut method is illustrated in Figure 7.15. For convenience we shall initially assume that it is the absolute tolerances that are being considered as fixed. Therefore the size of R_T is invariant whatever nominal design we consider. If, for whatever reason, a circuit described by the point S in parameter space is simulated and found to fail the specifications, then we know immediately that the nominal worst-case design *cannot* lie inside a rectangle having the same shape as R_T and centred on the fail point S. This rectangle is called a 'cut' because that region can be removed from consideration as a possible location for the nominal design. That the cut region has exactly the same shape as the tolerance region R_T can be appreciated by imagining an experiment in which R_T is moved around S in such a way that S lies just outside R_T (Figure 7.16). The assumption of fixed absolute tolerances allows us to identify the cut region associated with S easily

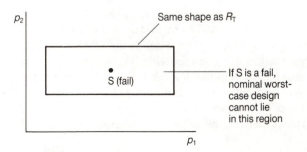

Figure 7.15
Identification of a cut region.

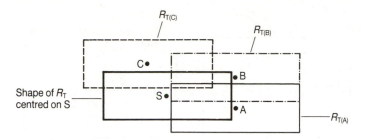

Figure 7.16
A fail point (S) lies at the centre of an area, having the shape of R_T, within which a worst-case nominal cannot lie.

and explicitly. The essence of the cut method is the gradual accumulation of cuts (i.e. areas of impossible nominal designs); at the same time information is gathered concerning potentially successful regions in which a nominal design can be located. In the discussion which follows we shall assume fixed absolute tolerances.*

The cut method is *iterative*, just like the centres of gravity method of design centring. Each iteration, the Jth (say), begins with the selection of a new trial nominal design P_J (using a strategy to be described), and the initiation of a Monte Carlo analysis. We say *initiation* because, as soon as a fail point (like S in Figure 7.15) is encountered, the Monte Carlo analysis is halted. The outcome of this exploration of parameter space within R_{T_J} is twofold, and provides the information necessary for any following iteration:

(1)   The single fail sample enables the corresponding cut region to be established as shown in Figure 7.17. This region is then amalgamated with all previously established cuts to identify the area in parameter space within which a successful nominal design cannot exist (Figure 7.17).

(2)   The pass samples generated before the single fail collectively indicate how successful the location of a nominal design nearby might be; the more pass samples encountered, the more likely it is that a successful nominal design will be found nearby. For example, if the Monte Carlo analysis runs to 300 samples before encountering a fail sample, it is easy to appreciate that only a small shift in the nominal design may be needed to achieve 100% yield.

* For the case of fixed relative tolerances, the cut region associated with a fail point S is slightly more difficult to compute. As before the cut region is such that if a point in it were to form a nominal (design) then the associated tolerance region would contain the fail point S, and therefore not be able to sustain a worst-case design. The difficulty now is that the size of the tolerance region R_T depends on its centre. One way round this difficulty is *not* to compute the cut region explicitly, but to note the position of the fail point S. Then for any point that is a candidate for worst-case nominal a simple test can be performed to check if its tolerance region contains S.

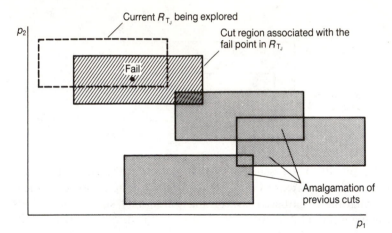

Figure 7.17

Incrementation of the cut region following the discovery of a new fail point.

Thus, the primary object of the exploration is not to estimate the yield associated with the current nominal design. Nevertheless, if the Monte Carlo analysis associated with an iteration does not discover a fail point within its first (say) 300 samples, then the designer may decide that, to a satisfactory approximation, the corresponding nominal design is an acceptable worst-case design, and the procedure halted.

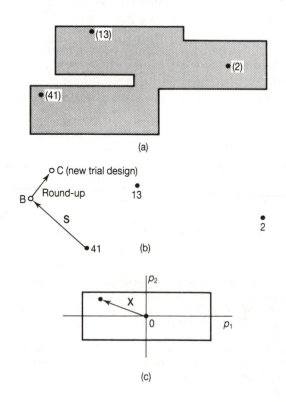

Figure 7.18

(a) The (shaded) amalgamated cut regions, the nominal designs (●) from which they were generated and, in parentheses, the number of pass samples encountered before the fail sample. (b) Generation of a new trial design. (c) Generation of a random vector within $R_T(0)$.

The strategy for selecting a new trial nominal design is based on two factors. First, of course, we know it cannot lie within the continuously updated cut region (Figure 7.17). Second, it must take account of the information described in (2) above, and select a trial nominal which, on the basis of the previous Monte Carlo analyses, appears to offer potential. The selection method is illustrated in Figure 7.18. Pertinent information relating to earlier iterations is shown in (a). Against each trial nominal is indicated the number (n) of pass samples that were generated before the fail sample was encountered. The nominal with the largest value of n is identified; although we know it is an unsatisfactory location for a nominal, a point nearby might nevertheless offer the best chance of finding a worst-case nominal. The way in which the 'nearby' point is selected is illustrated in Figure 7.18(b). Basically, a random vector $\mathbf{S}$ is added to the previous nominal design having the largest value of n, and the result (B in the figure) rounded to the nearest member (C) of the preferred (e.g. E12) current set of possible nominal designs. This point (C) is then the new trial nominal design. The generation of the random vector $\mathbf{S}$ is straightforward. First, a vector $\mathbf{X}$ is randomly generated within a tolerance region $R_{T(0)}$ having the origin as its centre (Figure 7.18(c)). This vector $\mathbf{X}$ is then magnified by a constant α to give $\mathbf{S}$:

$$\mathbf{S} = \alpha\mathbf{X}$$

In practice a value of α between 1.5 and 3.0 was found to give favourable results (Wehrhahn, 1984).

7.5.1 A practical implementation

An extensively tested and routinely used (Philips Te Ka De, Nurnberg) worst-case design technique devised by Wehrhahn (1982) employs the cut method described in outline above, and additionally uses correlated sampling. First, a tolerance region $R_{T(0)}$ is created with the origin of parameter space as its nominal, and sample points generated randomly according to a uniform distribution. We refer to this list of points as a_i; each point, of course, is described by the same number of coordinates as the parameter space has dimensions. Then, when generating samples for any real tolerance region it is only a straightforward matter to obtain, by a simple scaling and translation, the point b_i corresponding to a_i (Figure 7.19). In this way the relative spatial location of samples within each tolerance region is identical. This is essentially another instance of correlated sampling. It is always a sound strategy to employ the scheme whenever two situations with random elements are being

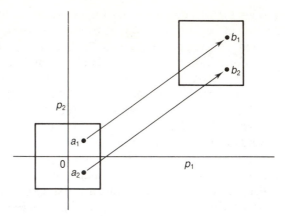

Figure 7.19
Scheme for generating random points for testing proposed worst-case solutions in the cut method.

considered. For the case of fixed absolute tolerances,* this approach involves just the addition of the coordinates $p_1^0 p_2^0 \ldots p_K^0$ of the nominal point P^0 to the corresponding coordinates of a_i.

For a given tolerance region, the order in which the sample points are analysed need not be the same for each iteration. Indeed, computational savings can be achieved by changing the order so that the 'most likely fail' points are analysed first. The way in which these points are selected is straightforward. As we have seen, a record is made of the first fail point encountered for each iteration. All that is necessary in a new tolerance region is first to analyse those (translated) points previously associated with a fail sample. Only when these points are exhausted are the remaining points analysed and tested.

Before the cut method is applied, the designer will decide upon the maximum number of samples to be tested within a single tolerance region such that, if no fail point is encountered, it is assumed that a satisfactory worst-case design has been achieved, and the procedure halted. This number is typically 300 or 400.

7.6   Circuit examples and results

7.6.1   The cut method

The first example of the application of Wehrhahn's cut method is shown in Figure 7.20. It involves a third-order RC-active low-pass filter; the resistors and capacitors have 1% tolerances and their nominals are

---

* For the case where the relative tolerances are fixed, each coordinate of a_i has initially to be scaled by the ratio of the tolerances associated with P^0 to those of $R_{T(0)}$. This then is added to the coordinates of P_j^0. If t_j is the tolerance of the jth component, for the nominal P^0 and t_j' is the corresponding tolerance of $R_{T(0)}$, then b_i^j the jth coordinate of b_i can be obtained from a_i^j the jth coordinate of a_i as

$$b_i^j = a_i^j \times t_j/t_j' + p_j^0$$

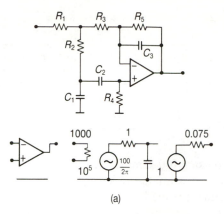

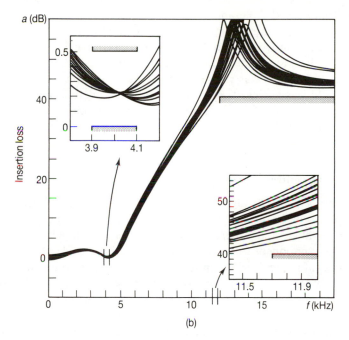

(b)

Figure 7.20
Third-order active filter example.

selected from the E96 series. The opamp's open-loop gain and its input and output resistances have a log-normal distribution which is fully accounted for in the sampling process, but these parameters have fixed nominals and cannot therefore be 'designed'. Figure 7.20(b) shows the insertion loss specifications, as well as the responses associated with 20 sample points selected from the Monte Carlo analysis associated with the final design. Details of the initial and final designs are shown in Table 7.1. A total of 21 cuts were required, involving a total of 1155 circuit analyses, though this performance should not be taken as typical or as the basis of a numerical comparison with another method. It is interesting to note that, for the same design example, the centres of

Table 7.1 Details of the initial and final design for the cut method applied to a third-order RC-active filter.

Components	$R_1(k\Omega)$	$R_2(k\Omega)$	$R_3(k\Omega)$	$R_4(k\Omega)$	$R_5(k\Omega)$	$C_1(\mu F)$	$C_2(\mu F)$	$C_3(\mu F)$
Starting values	13.7	0.127	0.221	16.2	14.0	0.0158	0.0110	0.00562
Final values	14.0	0.121	0.232	16.2	14.0	0.0191	0.0110	0.00604

gravity method using 100-sample iterations required four iterations and achieved successive estimated yields of 50%, 79%, 89% and 90%.

Wehrhahn's cut method has also been applied to the eleventh-order equalizer shown in Figure 7.21(a). Again, the resistors and capacitors have 1% tolerances and their nominals are selected from the E96 series, and the opamps have fixed tolerances and nominals. Figure 7.21(b) shows the specifications on the group delay and insertion loss, as well as the nominal performance of the final worst-case design. A total of 101 cuts was required, involving 1346 circuit analyses, and 100% yield was achieved. By contrast, the CoG method applied to the same problem took four iterations involving 400 analyses to reach an estimated yield of 95%.

7.6.2   The centres of gravity method

The centres of gravity algorithm has been applied to many circuits; two examples will be given here. Figure 7.22(a) shows the circuit of a high-pass filter, and Figure 7.22(b) the specifications on its insertion loss. The component tolerances were assumed to be 5% of the original nominal values, and the component pdfs were taken to be multivariate uniform with no correlation between components. Figure 7.23(a) shows the result of applying the CoG method, using 100-sample Monte Carlo analyses and employing the common points scheme. The number of fresh analyses required at each iteration, as well as the confidence of the correct ranking of yield estimates, is shown. In order to obtain confirmation of the result, 500-sample Monte Carlo analyses were used to check the initial and final yield estimates. It is seen that a useful increase in manufacturing yield was obtained in five iterations, at a not unreasonable cost in terms of circuit analyses.

To illustrate other aspects of design centring, the same filter was now assumed to be constructed from components having Gaussian pdfs. Such a situation cannot be handled by the common points scheme, so correlated sampling was employed in the application of the CoG algorithm; again, 100-sample Monte Carlo analyses were used. The result is shown in Figure 7.23(b). It is interesting, however, to exploit the

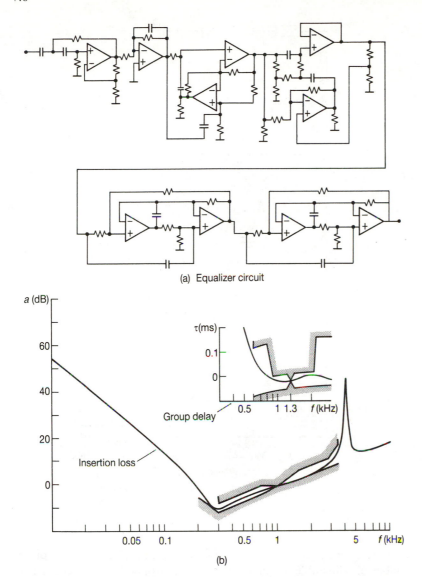

Figure 7.21
Insertion loss and group delay
response and performance
constraints.

computational advantage of the common points scheme by assuming a
multivariate *uniform* pdf for the components, carrying out four iterations of the CoG algorithm, and then switching to a Monte Carlo
analysis based on the Gaussian pdfs for the final check on yield. The
difference between the final yields (4%) in the two cases suggests that
some strategy of changing between correlated sampling and the
common points scheme may be advantageous.

To demonstrate that the centres of gravity algorithm is apparently insensitive to the dimensionality of a circuit, in the sense that

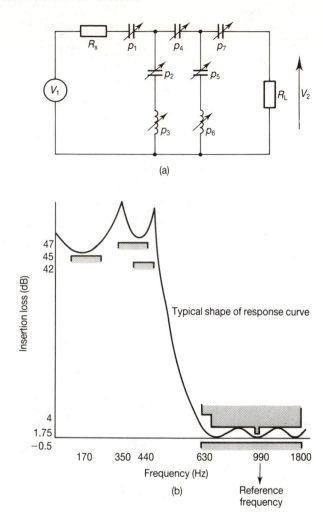

Figure 7.22

Passive high-pass filter.   (a) Circuit diagram. Arrows indicate toleranced components. Insertion loss is 20 log $|V_2(jw)/V_1(jw)|$.   (b) Performance requirements. Frequencies tested: 170, 350, 440, 630, 650, 720, 740, 760, 940, 1040, 1800 Hz. For initial parameter values, see Pinel and Roberts (1972).

the number of circuit analyses required is essentially independent of the number of adjustable nominal values, it was applied to a 43-coefficient integrated transversal filter employing charge-coupled devices. This type of filter has already been encountered in Chapter 1 (Figures 1.14 to 1.16): its coefficients are determined by integrated capacitors which, as a consequence of the fabrication process, are subject to statistical variation. The outcome of a design centring exercise on this filter is shown in Figure 7.24. Later, Knauer and Pfleiderer (1982) tested the CoG method by fabricating a large number of their initial and final designs, and obtained very good agreement between predicted and measured yield increases.

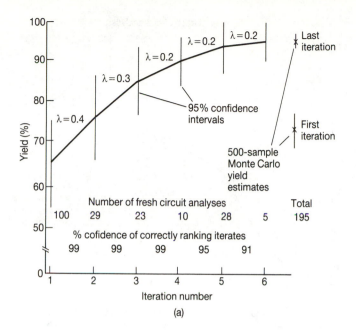

(a)

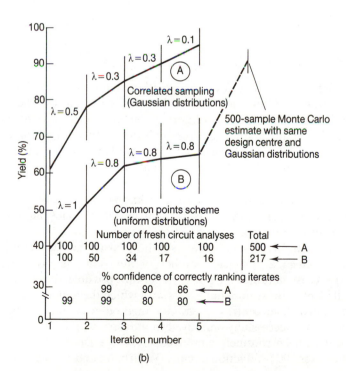

(b)

Figure 7.23

(a) Yield trajectory for high-pass filter assuming 5% tolerances and uniform distributions. (b) Yield trajectories for high-pass filter, assuming 10% tolerances and both Gaussian (A) and uniform (B) distributions.

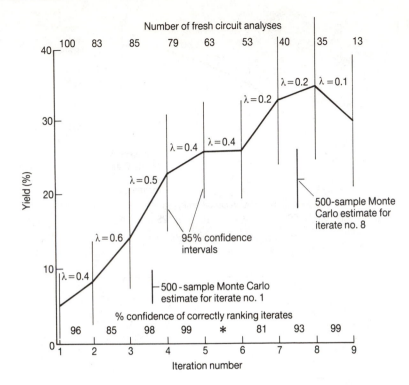

Figure 7.24
Yield trajectory obtained by
application of the centres of gravity
design centring algorithm to a 43-
coefficient transversal filter. *No
apparent increase in yield between
iterations 5 and 6.

7.7 Conclusions

The majority of tolerance design algorithms reported in the literature
are statistical in nature and based on Monte Carlo analysis. The reader
may well ask why the centres of gravity and cut methods were chosen
for inclusion in this chapter. The reason is that both methods have been
relatively successful, and are easy to understand and implement.

The centres of gravity method has been shown to give useful
increases in yield, but it must be noted that it will not necessarily
achieve a *global* yield maximum. Little is known theoretically about its
convergence properties (Styblinski, 1986), despite the fact that convergence problems rarely occur in practice.

Theoretically, if a worst-case solution exists, the cut method will
lead to it given a sufficient number of iterations; in other words,
convergence is guaranteed. If 100% yield is not possible, the cut method
will still identify those nominal points with which the highest yields may
be associated. The centres of gravity method was one of the first
methods to be successfully employed in industry and, by 1984, the cut
method had been routinely employed in Te Ka De Nurnberg in the
design of over 200 production circuits (Wehrhahn and Spence, 1984).

7.8 Appendix: Procedure for computing the confidence of correct ranking when employing the correlated sampling scheme

Let $P_1, P_2, \ldots, P_N$ represent the random circuits of the first Monte Carlo analysis and $Q_1, Q_2, \ldots, Q_N$ of the second Monte Carlo analysis. With each sample circuit we can associate a result 1 or 0 reflecting whether the random circuit was a pass or fail. Let us denote these results by the vector $\mathbf{u}_1, \mathbf{u}_2, \ldots, \mathbf{u}_N$ for the first analysis and $\mathbf{v}_1, \mathbf{v}_2, \ldots, \mathbf{v}_N$ for the second analysis. That is, $\mathbf{u}_1 = 1$ if the circuit P_1 is a pass and 0 if P_1 is a fail; similarly for $\mathbf{v}_1$ and Q_1. We now define quantities n_{11}, n_{00}, n_{10} and n_{01} in the expressions below.

$$n_{11} = \sum_{j=1}^{N} \mathbf{u}_j \cdot \mathbf{v}_j$$

$$n_{00} = \sum_{j=1}^{N} (1 - \mathbf{u}_j)(1 - \mathbf{v}_j)$$

$$n_{10} = \sum_{j=1}^{N} \mathbf{u}_j(1 - \mathbf{v}_j)$$

$$n_{01} = \sum_{j=1}^{N} (1 - \mathbf{u}_j)\mathbf{v}_j$$

These could quite easily be computed from the results of the Monte Carlo analysis.

Then clearly

$$\hat{Y}_1 = \frac{n_{11} + n_{10}}{N} \qquad \hat{Y}_2 = \frac{n_{11} + n_{01}}{N}$$

$$\widehat{\Delta Y} = \hat{Y}_2 - \hat{Y}_1 = \frac{n_{01} - n_{10}}{N}$$

As before

$$\widehat{\text{var}}(\hat{Y}_1) = \hat{Y}_1(1 - \hat{Y}_1)/N$$

and

$$\widehat{\text{var}}(\hat{Y}_2) = \hat{Y}_2(1 - \hat{Y}_2)/N$$

The formula for covariance between $\hat{Y}_1$ and $\hat{Y}_2$ is

$$\widehat{\text{cov}}(\hat{Y}_1, \hat{Y}_2) = (n_{11} n_{00} - n_{01} n_{10})/N^3$$

$$\widehat{\text{var}}(\widehat{\Delta Y}) = \text{var}(\hat{Y}_1) + \text{var}(\hat{Y}_2) - 2\widehat{\text{cov}}(\hat{Y}_1, \hat{Y}_2)$$

The confidence of correct ranking is now given by the formula

$$1/2 + \text{erf}\left(\frac{\sqrt{\widehat{\text{var}}(\widehat{\Delta Y})}}{|\widehat{\Delta Y}|}\right)$$

and can therefore be computed numerically.

CHAPTER 8

Methods using Yield Gradients

OBJECTIVES

If the effect of small variations in nominal component values on the manufacturing yield of a circuit can be estimated, such **yield gradient** information can be used to enhance the automatic and systematic optimization of the circuit. This chapter describes methods of obtaining yield gradient information and the way in which the resulting information can be utilized to optimize the tolerance properties of a mass-produced circuit. The methods of tolerance design discussed are of greater sophistication than the simple but robust methods for yield maximization described in Chapter 7. They also have broader application in the sense that component tolerances as well as nominal values are considered to be adjustable. Specifically, two methods are described which, in addition to estimating yield, also estimate derivatives (gradients) of yield with respect to component nominals and tolerances, and make use of these in an optimization procedure. The problems attending such an approach to tolerance design are discussed.

8.1   Use of yield gradients in tolerance design

It was shown in Chapter 5 that the results of the same Monte Carlo analysis that was used to estimate yield may also be employed to estimate the gradients (previously called the derivatives) of yield with respect to component nominals and tolerances. This additional estimation does not require any further pseudo-random circuit analyses, and hence incurs minimal extra cost. This chapter considers tolerance design methods which make use of such yield gradient information. Two basic approaches have been pursued by different researchers and are outlined below.

Perhaps the most obvious strategy, and one we shall call the *general approach*, is to execute an iterative procedure where, at each iteration and for a particular set of nominals and tolerances (i.e. a trial circuit design), the yield and its gradients are estimated via Monte Carlo analysis, and are subsequently used to choose a new trial design. When compared with the centres of gravity method of design centring, this approach simply entails replacing the CoG algorithm for choosing new nominals with a more sophisticated method employing yield gradients (see Figure 8.1). The simplest gradient-based optimization procedure is that of steepest ascent. However, since the Monte Carlo analysis can also provide estimates of second-order yield gradients, a more effective gradient-based method (e.g. Fletcher and Powell, 1963) may be used.

In the second approach, yield and possibly its gradients are first estimated for a particular nominal point in parameter space; that is, for a given trial design. These estimates are then substituted into an approximating analytic relationship referred to as a **yield prediction formula** (YPF), which relates *yield* to *nominal points* in the vicinity of the trial design (Figure 8.2). Basically, a YPF allows yield estimates to be extrapolated to new nominal points in the vicinity of the trial nominal point for which the Monte Carlo analysis took place. The next

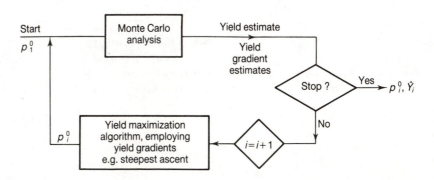

Figure 8.1

Outline description of a simple gradient-based yield maximization scheme.

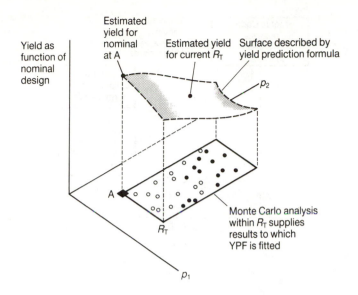

Figure 8.2

The basis of the yield prediction formula approach to tolerance design.

step in the YPF procedure is to apply gradient-based optimization – based on the YPF and *not* on further circuit analyses – and thereby (Figure 8.3) find a local optimum. This local optimum then becomes the new trial design: a new Monte Carlo analysis is then performed for this design, a fresh YPF is set up and the iterative process continues.

8.2   The problems addressed

Before discussing yield-gradient-based tolerance design methods in detail we shall indicate the various types of problem that can be addressed by these methods. In Chapter 7 two algorithms were described that dealt with a *single* problem, that of yield maximization. By contrast, the methods discussed in the present chapter have more general application and can, for example, simultaneously include yield maximization and cost reduction by tolerance assignment. In all cases an essential prerequisite is the definition of a *cost function* which is to be optimized. In yield maximization it is Y that is to be maximized, whereas in tolerance assignment it may be the average cost of manufacturing a successful circuit that must be minimized.

As an illustrative example we consider the tolerance assignment problem in which the nominal component values are fixed and the average cost of successful circuits is to be minimized by judicious choice of component tolerances. The cost of an individual component may reasonably be taken to vary inversely with its tolerance, so that

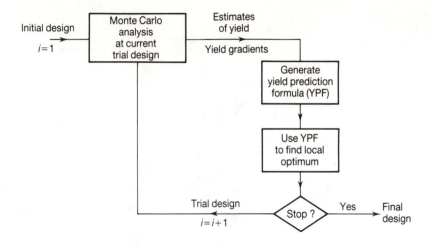

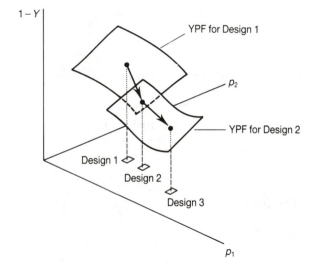

Figure 8.3
Iterative optimization procedure
based on the generation, at each
iteration, of a yield prediction
formula. Two iterations are shown:
at each, a Monte Carlo analysis is
required to generate the yield
prediction formula.

the combined cost of all the components in a circuit may be expressed
as

$$C_c = \sum_{i=1}^{K} \alpha_i + \frac{\beta_i}{t_i} \tag{8.1}$$

For the ith of the total of K components, α_i and β_i are constants and t_i
is the tolerance. An overall *unit cost* function, which is the average cost
of producing a circuit that meets the specifications, may therefore be of
the form

$$C_u = \frac{C_c}{Y} \tag{8.2}$$

simply because a fraction $1 - Y$ of the manufactured circuits must be discarded.* The objective of tolerance assignment is to minimize C_u in Equation 8.2 by suitable choice of component tolerances t_i.

The gradients of the unit cost function C_u with respect to tolerances will be of the form

$$\frac{\partial C_u}{\partial t_i} = \frac{Y\dfrac{\partial C_c}{\partial t_i} - C_c\dfrac{\partial Y}{\partial t_i}}{Y^2} \tag{8.3}$$

We note that both the expression for unit cost C_u (Equation 8.2) and its derivative (Equation 8.3) involve two types of quantity. On the one hand there are quantities such as yield and yield gradient which can only be *estimated* – and usually at *substantial cost* – perhaps directly from a Monte Carlo analysis or indirectly via a yield prediction formula. On the other hand quantities such as the gradient of component cost ($\partial C_c/\partial t_i$) can easily be found from an *analytic* cost function (Equation 8.1). Other tolerance design problems will in general involve these two types of quantity.

8.3   Method of parametric sampling

In Section 8.1 we described and illustrated (Figure 8.1) in outline a 'general approach' to tolerance design in which, following a Monte Carlo sampling within the tolerance region, the estimated yield and yield gradients are employed within some standard optimization method such as steepest ascent. Such a method was in fact implemented by Batalov *et al.* (1978) with encouraging results. However, because their approach is unnecessarily inefficient (in that it required a full Monte Carlo analysis at each iteration without re-using much of the information generated in previous Monte Carlo analyses), we choose instead to present a more recent method, developed by Singhal and Pinel (1981), called the method of parametric sampling.

Basically, the aim of the method of parametric sampling is to avoid the high cost associated with iterative schemes in which a Monte Carlo analysis is required at each iteration. In outline, the solution proposed by Singhal and Pinel was that successive iterations should draw their samples from a 'pool' generated earlier in a *single* sampling within an exploratory region R_E of parameter space. Since some of the samples in the 'pool' may be used in more than one iteration there is a

---

* For simplicity of illustration, no account is taken of the cost of measuring the circuits to see which ones should be discarded. The measurement cost may, however, easily be incorporated in the cost function.

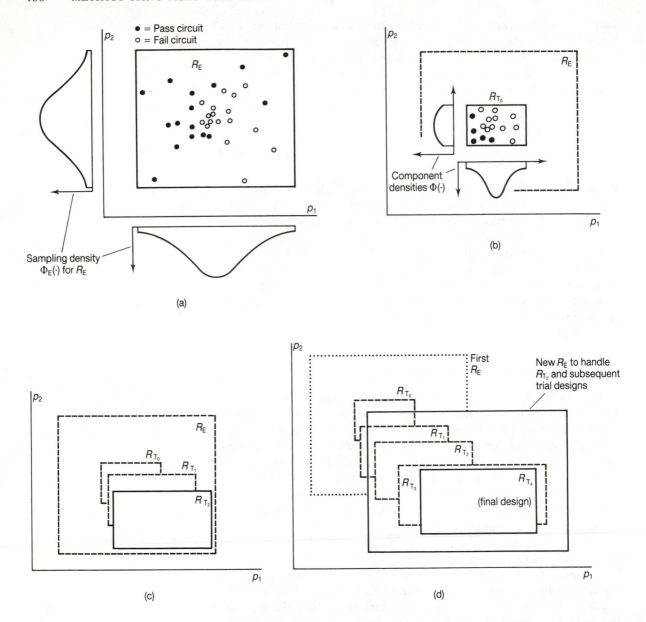

Figure 8.4
The method of parametric sampling. (a) Random selection, analysis and testing of a 'pool' of circuits within a region of exploration. (b) Those samples appearing in the R_{T_0} 'window' are treated as conventional Monte Carlo samples. (c) The iterative scheme terminates within R_E, with the final, optimized design described by the tolerance region R_{T_2}. In the course of the optimization some of the 'pool' of samples will have been used more than once, some not at all. (d) Movement of the trial designs may be such that a new 'pool' of samples is required.

resemblance to the common points scheme (Section 7.4), though without any restriction to uniform component pdfs.

To illustrate the method of parametric sampling we consider the general tolerance assignment problem in which a cost function (Example 8.2, for example) involving tolerances and yield is to be minimized by judicious choice of component tolerances and nominal values. The first step (Figure 8.4(a)) in the procedure is the random generation, within a region of exploration (R_E) somewhat larger than R_T, of a large number (e.g. 1000) of sample points. The sampling takes place according to a pdf $\phi_E(.)$ which is typically multivariate Gaussian but which is not in general related to the actual component pdfs. A circuit analysis is then carried out for each sample point, and each is classified as pass or fail by reference to the specifications. In the next step (Figure 8.4(b)) the initial tolerance region R_{T_0} corresponding to the initial circuit design is located within R_E. The (already generated) samples that fall within R_{T_0} are then treated as conventional Monte Carlo samples from which the yield and its gradients can be estimated at negligible additional cost. Account must of course be taken of the fact that the actual component pdfs $\phi(.)$ normally differ from those ($\phi_E(.)$) associated with R_E. On the basis of the estimated yield and yield gradients (found by a method to be described below) a new nominal point and tolerances are chosen and the new tolerance region R_{T_1} located. The process is then repeated (Figure 8.4(c)) either until a local minimum of the cost function is deemed to have been achieved or until (Figure 8.4(d)) a new 'pool' of samples within a new R_E is required to allow progress of the iterative scheme.

8.3.1 Parametric estimators

In a conventional Monte Carlo analysis, pseudo-random points are generated according to the component pdf $\phi(.)$. The yield is then estimated as the ratio of the passing points to the total number (N) of points. In Chapter 4 the yield estimate was expressed as

$$\hat{Y} = \frac{1}{N} \sum_{j=1}^{N} g(P_j) \tag{8.4}$$

where $g(P)$ is the testing function and has a value of 1 or 0 according to whether the point P_j represents a pass or fail circuit respectively. But in the method of parametric sampling the pdf $\phi_E(.)$ according to which samples are *generated within* R_E is different from the component pdf. Under these circumstances each pass point (for which $g(.)$ would normally be unity) is now assigned a weight equal to the ratio of the value of the component pdf to the sampling pdf at that point. Thus, the relevant yield estimator is

$$\hat{Y} = \frac{1}{N} \sum_{j=1}^{N} g(P_j) \frac{\phi(P_j)}{\phi_E(P_j)} \tag{8.5}$$

As a reminder we note that summation 8.4 is an estimate of the integral defining yield:

$$Y = \int_{R_E} \ldots \int \{g(P)\} \, \phi(P) \, dP \tag{8.6}$$

This integral (Equation 8.6) may be rewritten as

$$Y = \int_{R_E} \ldots \int \left\{ g(P) \frac{\phi(P)}{\phi_E(P)} \right\} \phi_E(P) \, dP \tag{8.7}$$

Both integrals (Equations 8.6 and 8.7) are the expected value (average) of the term in curly brackets with respect to a pdf: in both cases the Monte Carlo estimate is obtained by sampling according to the particular pdf (ϕ in Equation 8.6 and ϕ_E in Equation 8.7) and finding the mean of the function in curly brackets evaluated for each of the sample points. The modification of the integrand defining the yield, as in Expression 8.7, allows the pdf from which the random samples are taken to differ from that which (physically) applies to the component parameters. Such a modification is referred to as the *importance sampling relationship*; both the methods of tolerance design (parametric sampling and the yield prediction formula) described in this chapter are based on it.

Having derived estimators for yield we now proceed to obtain an estimator for yield *gradient* relevant to the parametric sampling scheme of Singhal and Pinel. First, however, it is useful to draw attention more explicitly to the nature of the adjustable parameters.

The parametric sampling optimization scheme involves a number of different trial designs: each will be characterized by the same form of component pdf $\phi(.)$ but by different nominals and sets of tolerances. For example, for a multivariate Gaussian distribution of the component values, the means will correspond to nominal component values and the *i*th standard deviation σ_i may be taken to be related to the tolerance of the *i*th component according to $t_i = 3\sigma_i$. In a more general situation correlations may exist between different component values so that a variance–covariance matrix must be considered. At this juncture *all* these parameters (nominals, tolerances, etc.) for the N components are denoted by the vector $\mathbf{X}$ where

$\mathbf{X} = x_1, x_2, \ldots, x_i, \ldots, x_N$. The component pdf will now be denoted by $\phi(P, x)$ where, as before, P is a general point in parameter space, and the inclusion of $\mathbf{X}$ reflects the dependence of the pdf on nominals, tolerances and correlations.

With the notation just introduced, the integral (Equation 8.7) defining yield becomes

$$Y(\mathbf{X}) = \int \ldots \int_{R_E} \left\{ g(P) \frac{\phi(P_i \mathbf{X})}{\phi_E(P)} \right\} \phi_E(P) dP \qquad (8.8)$$

Differentiation of Equation 8.8 with respect to the parameters $\mathbf{X}$ then yields

$$\frac{\partial Y}{\partial \mathbf{X}} = \int \ldots \int_{R_E} \left\{ \frac{g(P)}{\phi_E(P)} \cdot \frac{\partial \phi(P_i \mathbf{X})}{\partial x_i} \right\} \phi_E(P) dP \qquad (8.9)$$

This equation not only expresses the yield gradient but has the same form as the integrals in Equations 8.6, 8.7 and 8.8. Therefore the corresponding yield gradient *estimator* may, by comparison with Equations 8.4 and 8.5, easily be expressed as

$$\frac{\widehat{\partial Y}}{\partial \mathbf{X}_i} = \frac{1}{N} \sum_{j=1}^{N} \frac{g(P_j)}{\phi_E(P_j)} \cdot \frac{\partial \phi(P_j, \mathbf{X}_i)}{\partial \mathbf{X}_i} \qquad (8.10)$$

If the ratio of pdfs $\phi(P, \mathbf{X})/\phi_E(P)$ is denoted by $w(P, \mathbf{X})$ to characterize the 'weight' of the point P, Equation 8.10 may more conveniently be expressed as

$$\frac{\widehat{\partial Y}}{\partial \mathbf{X}_i} = \frac{1}{N} \sum_{j=1}^{N} g(P_j) \frac{\partial w(P_j, \mathbf{X})}{\partial \mathbf{X}_i} \qquad (8.11)$$

In terms of the same notation, the yield estimator (Equation 8.5) becomes

$$\hat{Y} = \frac{1}{N} \sum_{j=1}^{N} g(P_j) w(P_j, \mathbf{X}) \qquad (8.12)$$

Estimators such as 8.11 and 8.12 are referred to as **parametric estimators**. A similar development leads to estimators for second-order yield gradients:

$$\frac{\widehat{\partial^2 Y}}{\partial \mathbf{X}_i \partial \mathbf{X}_j} = \frac{1}{N} \sum_{j=1}^{N} g(P_j) \frac{\partial^2 w(P_j, \mathbf{X})}{\partial \mathbf{X}_i \partial \mathbf{X}_j} \qquad (8.13)$$

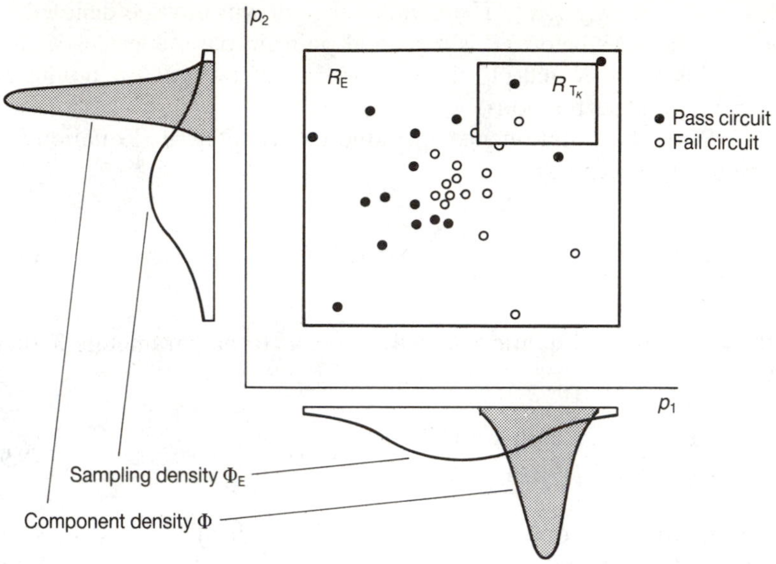

Figure 8.5
Illustration of the need to update the database. After a number of iterations the component pdf may be severely disjoint with respect to the sampling pdf, so that few of the original circuit samples are relevant to the current tolerance region. As a consequence, the yield and yield derivative estimates may be inaccurate.

8.3.2   Database updating

Formulae 8.10 to 8.13 show how yield and yield gradients may be estimated via a Monte Carlo analysis in which the applicable component pdf and the pseudo-random sampling pdf are different. Essentially this involves the use of weights which are the ratios of the values of the two pdfs at the relevant points in component space. One advantage of the parametric sampling method is that the result of some circuit analyses may be used more than once in successive iterations (though, of course, some may not be used at all). An attendant disadvantage, however, is that when the component pdf 'moves away' substantially from the sampling pdf (as it may well do after a number of iterations as shown in Figure 8.5), the number of points available in the current tolerance region may be so small as to lead to an insufficiently accurate estimate of yield. In this case consideration must be given to the generation of additional data points, a process we shall term **database updating**. Three different strategies have been proposed for making the decisions concerning *when* to generate new samples and *how* to update the database of sample points.

STRATEGY A

In the first strategy the weights are calculated for all the points in the existing database which are relevant to the current (Kth) iteration; that is, for all sample points falling within R_{T_K}. Those points having a weight greater than a constant E (typically $E = 0.2$) are then regarded as meaningful points for the current iteration. Let there be n_K such

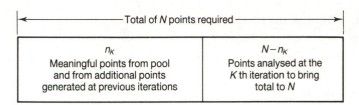

Figure 8.6
A strategy for updating the database of sample circuits.

points; i.e. $w(p, x_K) > E$ for n_K values of P. Again, x_K denotes all the component nominals and tolerances selected at the Kth iteration. If n_K is less than a minimum number (N) of samples deemed to be adequate for estimation purposes (typically $N = 100$), then additional sample points are generated according to the original sampling distribution $\phi_E(p)$ and their weights are calculated according to

$$w(P_i, \mathbf{X}_K) = \frac{\phi(P, \mathbf{X}_K)}{\phi_E(P, \mathbf{X}_K)}$$

as before. Only if the weight of a point is greater than E is it accepted as a meaningful point for the current iteration and the corresponding circuit then analysed. The process of sample point generation and weight checking is continued until the total number of meaningful points reaches the predetermined value N. The process can be illustrated by Figure 8.6. In this particular strategy, the database may well be updated at every iteration.

STRATEGY B

A second strategy is to destroy the original database after a predetermined number of iterations and construct a fresh database centred about the current trial design. In general this strategy leads to a greater accuracy of yield estimation than does the first strategy, but incurs a greater computational cost. The criteria for deciding *when* to create an entirely new database are discussed below.

STRATEGY C

A third strategy is to generate a new database but *not* to destroy the original one. Instead, a 'pooled estimate' of yield is obtained from both old and new sample points. To explain this strategy we assume that a number (L) of databases have earlier been generated at previous iterations. By using these L databases, L different parametric yield estimates and L sets of estimated yield gradients can be obtained for the current trial design. If we denote by $\hat{Y}_j(x_K)$ the parametric yield estimate obtained for the current (Kth) iteration using the jth database, then a *pooled estimate* of yield for the Kth iteration can be defined as a weighted sum of the L (generally different) estimates:

$$\hat{Y}_K = \sum_{j=1}^{L} \theta_j Y_j(\mathbf{X}_K) \tag{8.14}$$

The weights θ_j satisfy the condition

$$1 = \sum_{j=1}^{L} \theta_j$$

The actual values of θ_j are chosen to minimize the variance of $\hat{Y}$. Singhal and Pinel have suggested that the weights be selected according to

$$\theta_j = \frac{1}{\text{var}\left[\hat{Y}_j(\mathbf{X}_K)\right]\left\{\sum_{j=1}^{L} \dfrac{1}{\text{var}\left[\hat{Y}_j(\mathbf{X}_K)\right]}\right\}} \tag{8.15}$$

where $\text{var}\left[\hat{Y}_j(\mathbf{X}_K)\right]$ is the variance of the parametric yield estimator $Y_j(\mathbf{X}_K)$ and is given by

$$\text{var}\left[\hat{Y}_j(\mathbf{X}_K)\right] = \frac{1}{N-1}\left[\frac{1}{N}\left\{\sum_{i=1}^{N} g(P_i)w(P_i, \mathbf{X}_K)\right\} - \hat{Y}_j^2(\mathbf{X}_K)\right] \tag{8.16}$$

DATABASE UPDATE

We now turn to the question, relevant to the second and third strategies described above, as to *when* the database of sample points should be updated. Essentially, the answer is to place *bounds* on the region in which a current tolerance region (R_{T_K}) may lie, these bounds being so chosen as to ensure a sufficient density of sample points in R_{T_K} with which to estimate yield and its gradients with acceptable accuracy. It is when any of these bounds is about to be violated that the database needs updating. The details of a suitable scheme are described below in the remainder of this subsection.

Assume for the purpose of illustration that all *component* pdfs are Gaussian with mean (nominal value) μ_i and standard deviation σ_i where, as usual, i identifies the ith component. Let the sampling pdfs for the Lth database also be Gaussian with mean μ_{L_i} and standard deviation σ_{L_i}. We assume that the means of the two pdfs are identical* (Figure 8.7)

$$\mu_{L_i} = \mu_i \tag{8.17}$$

---

* Knowledge of the likely search direction (i.e. the direction of movement of R_T) may suggest that the sampling mean μ_{L_i} be biased in that direction.

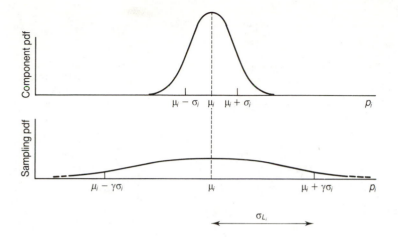

Figure 8.7
The sampling pdf is broader than the initial component pdf to allow for movement and possibly expansion of R_T.

and that, to allow optimization to proceed the sampling pdf is the 'broader' of the two by an inflation factor γ (Figure 8.7):

$$\sigma_{L_i} = \gamma \sigma_i \tag{8.18}$$

Experience suggests a value of about 4 for γ. Using the means and standard deviations of Equations 8.17 and 8.18 the Lth database is now generated.

The optimization algorithm will now be initiated in order to carry out the tolerance design task. Recalling Section 8.2 the objective may, for example, be to minimize unit circuit cost by suitable choice of nominals and tolerances, in which case the optimization objective can be stated as

$$\underset{X}{\text{Minimize}}\ C_u(X, Y_L(X)) \tag{8.19}$$

The subscript L refers to the Lth database, and X the nominals and tolerances. However, to ensure an adequate density of sample points in any tolerance region we consider a *constrained* minimization problem in which the objective (8.19) remains the same but three sets of constraints are imposed. The first, illustrated in Figure 8.8, ensures that the weights $w(P, X)$ associated with sample points do not become too small by ensuring that the 'low probability extremes' of the component pdfs do not move outside those of the sampling pdf. Formally, this set of constraints can be expressed as

$$\mu_i' + \beta_2 \sigma_i' \leqslant \mu_{L_i}' + \beta_1 \sigma_{L_i}'$$
$$\mu_i' - \beta_2 \sigma_i' \geqslant \mu_{L_i}' - \beta_1 \sigma_{L_i}' \tag{8.20}$$

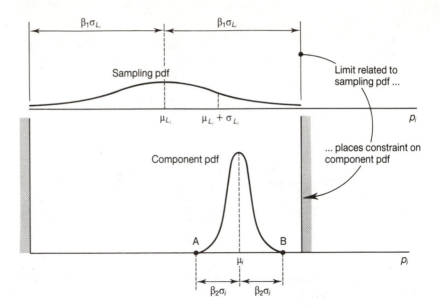

Figure 8.8
Constraints on the component pdf
are generated from details of the
sampling pdf. Points A and B must
lie within the shaded constraints.

where superscript primes draw attention to the fact that, during the optimization, component nominals and tolerances will have taken on new values. These constraints ensure that the $\pm\beta_2\sigma_i$ limits of the component pdf do not exceed the $\pm\beta_1\sigma_i$ limits of the exploratory sampling pdf. The constants β_1 and β_2 are typically chosen to be equal to 2.

The changes in means and variances from one iteration to the next are constrained according to

$$|\gamma_i\sigma_i - \sigma'_{L_i}| \leqslant \delta_{L_1}$$

$$\left|\frac{\mu_i - \mu'_{L_i}}{\mu'_{L_i}}\right| \leqslant \delta_{L_2} \tag{8.21}$$

If, during the constrained optimization, any of the constraints expressed by inequalities 8.20 and 8.21 are violated, then the database is updated by shifting the mean of the sampling pdf to the new trial design and by selecting its standard deviation according to Equation 8.16. The optimization then continues and is terminated only when a solution to the constrained optimization problem is found which does not violate any of the constraints expressed by inequalities 8.20 and 8.21. The values of the constants γ, δ_{L_1} and δ_{L_2} are reduced each time the database is updated since the tolerance design is typically converging and large changes in μ and σ are consequently no longer required.

8.3.3   Circuit examples

The performance of the parametric sampling method is illustrated here by results obtained for two tolerance design problems. The first example concerns a high-pass filter whose circuit diagram, nominal response and performance specifications are shown in Figure 8.9. Singhal and Pinel (1981) employed the cost model

$$C_u = \frac{C_A + \sum_{\text{capacitors}} \frac{1}{t_i} + \sum_{\text{inductors}} \frac{2}{t_i}}{Y} \qquad \textbf{(8.22)}$$

where C_A, representing the tolerance-independent fixed manufacturing costs, was taken to be 2.5, and where the tolerance-dependent costs of the capacitors and inductors were taken to be $1/t_i$ and $2/t_i$ respectively. The components were assumed to be independent and normally distributed (with truncation at $\pm 3\sigma$) except for the resistors modelling

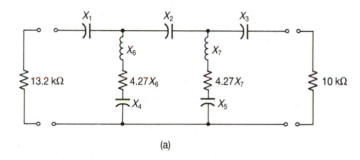

(a)

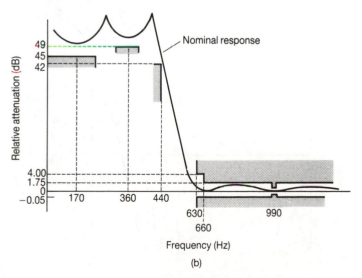

(b)

Figure 8.9
Illustration of the method of parametric sampling.   (a) Circuit diagram of high pass filter. (b) Nominal response and performance requirements of high pass filter.

Table 8.1   Results for the high-pass filter example. Inductance values are shown in henries and capacitors in nanofarads.

Component	Starting nominal values (μ)	Design cycle with three databases					
		100-point database		Additional 200 points		Additional 300 points	
		$\Delta\mu(\%)$	$t(\%)$	$\Delta\mu(\%)$	$t(\%)$	$\Delta\mu(\%)$	$t(\%)$
X_1	11.80	-1.80	9.25	-1.01	8.06	1.11	9.33
X_2	8.73	-1.50	7.25	0.92	7.64	1.43	8.00
X_3	10.45	3.83	8.92	6.77	11.05	6.52	9.39
X_4	39.04	-3.39	10.00	-3.22	9.34	-4.20	9.27
X_5	90.13	4.35	7.97	3.66	10.11	5.04	14.00
X_6	3.85	0.40	10.00	2.94	12.50	1.36	8.87
X_7	3.18	-3.57	10.00	-3.53	10.34	-2.93	11.85
Yield verification based on 300 samples		89.67		84.33		97.33	
Cost estimate		3.88		3.78		3.50	

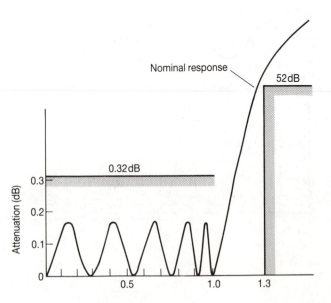

Figure 8.10
Eleventh order low-pass filter circuit example: circuit diagram, nominal response and performance specifications.

inductor losses: the latter were assumed to be completely correlated with, and linearly proportional to, the inductor values. The nominal values of the components corresponded to the mean values μ_i of the component distributions, while their fractional tolerances t_i were related to their standard deviations as $t_i = 3\sigma_i/\mu_i$. Starting from an initial design, optimization by the method of parametric sampling involved three databases of sample sizes 100, 200 and 300 and reduced the unit circuit cost C_u from 3.88 to 3.5 units. For each intermediate solution a 300-sample Monte Carlo analysis was performed for verification purposes. The results are summarized in Table 8.1.

The second tolerance design problem involved the eleventh-order low-pass filter shown in Figure 8.10. A cost function of the form

$$C_u = \left(\sum \frac{1}{t_i} \right) \bigg/ Y$$

was reduced from 44.3 to 7.48 with the first database of 3000 points and to 6.76 following the generation of the second database of 500 points.

8.4 A method using yield prediction formulae

A second approach to tolerance design involving yield gradients is based on yield prediction formulae. Such an approach was developed and incorporated in an interactive package called TACSY (tolerance analysis and design centring system) by Antreich and his co-workers at the Technical University of Munich. First, we give a brief example to illustrate the relevant concept.

A simple way to appreciate the concept of a yield prediction formula is as follows. We assume for simplicity that component tolerances are fixed, and that the component pdf $\phi(P, P^0)$ is centred about the nominal point P^0. Then a simple, though in practice not a very effective, yield prediction formula is the familiar Taylor series truncated after the second derivative:

$$Y(P^0) = Y(P_1^0) + \sum_{i=1}^{K} \Delta P_i^0 \frac{\partial Y}{\partial P_i^0} + \sum_{i=1}^{K} \sum_{j=1}^{K} \Delta P_i^0 \frac{\partial^2 Y}{\partial P_i^0 \partial P_j^0} \Delta P_j^0$$

where $P_1^0 = p_{11}^0, p_{12}^0, p_{13}^0, \ldots, p_{1K}^0$ is the nominal design at which yield has been estimated by Monte Carlo analysis, and P^0 is a general point in the vicinity. The change in nominal points is denoted by $\Delta P^0 = P^0 - P_1^0$. This yield prediction formula predicts the yield that will be obtained if the nominal point is moved from P_1^0 to another point P_2^0. If the problem addressed is that of yield maximization, then a point P_2^0 could be found such that the yield Y predicted by the above expression is maximized. In the tolerance design method under consideration a new Monte Carlo analysis would then be performed with P_2^0 as the nominal, the yield prediction formula would be updated, and the procedure would continue until terminated by the designer.

Following that brief introduction to the concept of a yield prediction formula we now examine the method in greater depth. We again assume, for purposes of illustration, that component tolerances are fixed. As before, a particular form of component pdf $\phi(P, P^0)$ will be considered, where P^0 denotes the nominal values of the components (i.e. the nominal design) which may be varied in order to maximize circuit yield. In the Antreich method a conventional Monte Carlo analysis is first performed using the component pdf $\phi(P, P_1^0)$ centred about a 'base point' which is the nominal point P_1^0: the subscript 1 refers to the *first* design (in other words, the first of a number of exploratory nominal points in parameter space). The results of this Monte Carlo analysis are then used to estimate both yield and yield gradients as described in Chapter 5. These estimates are then substituted into two formulae. The first formula, the Yield Prediction Formula (YPF), predicts the yield for any choice of the new nominal point in the vicinity of the base point:

$$\hat{Y}_{\mathrm{p}} = f(\hat{Y}_i, \hat{S}_i, P^0) \tag{8.23}$$

Here $\hat{Y}_1$ is the Monte Carlo yield estimate for the nominal point P_1^0, and $\hat{S}_1$ is the yield gradient estimate. P^0 represents the new nominal point whose associated yield is to be predicted. The manner in which the detailed formula is obtained is discussed in Section 8.6.

A second formula, the variance prediction formula, predicts the variance V_{p} of the yield estimate obtained from the yield prediction formula (8.23):

$$\hat{V}_{\mathrm{p}} = f^1(\hat{Y}_i, \hat{S}_i, P^0) \tag{8.24}$$

This variance prediction formula (VPF) provides an indication of the uncertainty associated with the estimates of yield computed using the YPF (Equation 8.23). The VPF permits the monitoring of the optimization procedure used to maximize yield, in a manner soon to be described.

As already stated, Antreich's method attempts to maximize yield as given by the YPF. Clearly, the accuracy of the YPF will be greatest in the proximity of the base point P_1^0 and will diminish as points further away are considered. Therefore, the search for a new nominal point which maximizes the yield should be constrained to the vicinity of P_1^0. The normalized magnitude of the deviation from the base point is a suitable measure of proximity to P_1^0, and will be constrained to be less than some constant. If P_2^0 is the new nominal being considered, and $\Delta P = P_2^0 - P_1^0$ is the corresponding change in nominal points, we shall require that the normalized distance ΔP satisfy the condition

$$\| \Delta P \| \triangleq \sqrt{(\Delta P^{\mathrm{T}} C^{-1} \Delta P)} \leqslant X_0 \qquad \textbf{(8.25)}$$

where C is the variance–covariance matrix of the component pdf ϕ and X_0 is a positive constant. For example, if $\phi(.)$ was a multivariate normal pdf with no correlation between the components, then C would be a diagonal matrix and, for different values of the constant X_0 (8.25), represents spheres concentric with the nominal point P_1^0. For the multivariate normal case, 8.25 represents equiprobable curves of the component pdf.

We are now in a position to consider the following constrained optimization problem:

$$\text{Maximize } Y_{\mathrm{p}}(\hat{Y}_1, \hat{S}_1, P^0) \qquad \textbf{(8.26)}$$

subject to

$$\| \Delta P \| = \sqrt{(\Delta P^{\mathrm{T}} C^{-1} \Delta P)} \leqslant X_0$$

By means of a suitable optimization algorithm, problem 8.26 is now solved *separately* for several successively increasing discrete choices of the constant X_0. At each relaxation of the constraint represented by X_0, a new nominal design is obtained; we shall denote them by $P_2^0, P_3^0, P_4^0, \ldots, P_n^0$. For each nominal design and its associated maximized yield, the variance given by the VPF is also computed. This procedure now allows the designer to be presented with two curves, one showing maximized yield versus X_0, and the other showing the square root of the corresponding variance. The curves shown in Figure 8.11 are typical of the early stages of such an optimization. In the examples shown, the predicted maximized yield increases with distance from the base point and, not unexpectedly, so does the predicted standard deviation, the square root of the predicted variance. The designer can now make a choice of X_0, striking some compromise between predicted yield and statistical uncertainty. If the two curves indicate no suitable compromise, then the number of Monte Carlo samples associated with the base point P_1^0 can be increased and the procedure repeated.

Having selected a value for the constant X_0, the new nominal design P_2^0 is obtained. This design now serves as the next base point about which a new Monte Carlo analysis is performed. Yield and yield gradients are estimated for the new base point P_2^0 and new yield and

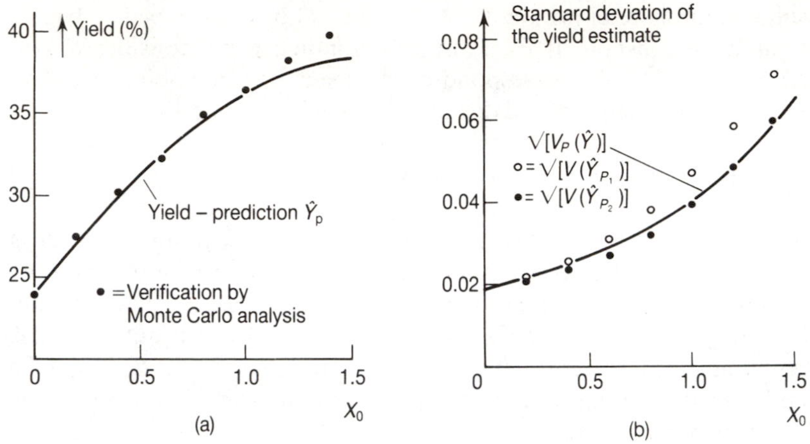

Figure 8.11
(a) Predicted yield, maximized for various values of normalized distance from base point X_0.
(b) Predicted variance against X_0.

variance prediction formulae are determined. The optimization procedure is repeated until terminated by the designer. The process is summarized by the flowchart shown in Figure 8.12.

The present section has presented only a simplified overview of the yield prediction formulae approach to tolerance design. A more detailed discussion is presented in Section 8.6.

8.4.1   Circuit example

The fifth-order switched-capacitor low-pass filter of Figure 8.13 is intended for pulse code modulation (PCM) applications. The nominal values of the capacitors constitute the 12 designable parameters of this circuit. The distributions of the capacitors are assumed to be normal with the initial nominal (mean) values shown in Figure 8.13(a). The standard deviations are taken to be a constant fraction (0.02) of the means, i.e. $\sigma_i = 0.02 P_i^0$. The authors of the method took the correlation coefficients between the random capacitance values to be zero. Since the functional accuracy of switched-capacitor circuits depends on the strong correlations between on-chip capacitances in MOS technology, the latter assumption renders this example of little value as a practical circuit design exercise: nevertheless, it is adequate for illustrating the power of the design method. The performance requirements and a typical acceptable circuit response are shown in Figure 8.13(b).

Using the nominal values shown in Figure 8.13(a), a Monte Carlo analysis with a sample size of 500 was performed, and the corresponding yield and variance prediction formulae were obtained. The constrained optimization problem (8.26) was then solved several times using progressively larger values of X_0, where X_0 is the distance from the initial nominal point.

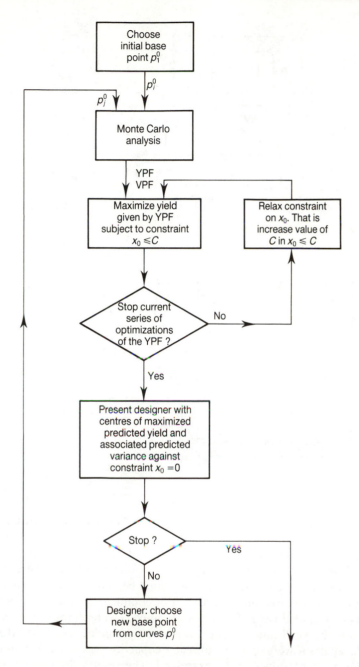

Figure 8.12
Overall yield maximization method using yield prediction formulae.

Figure 8.11 shows a plot of the optimized (maximized) yield as predicted by the YPF, and the corresponding variance as predicted by the VPF, for the different values of X_0. From visual examination of these two plots the designer made the empirical decision to choose a

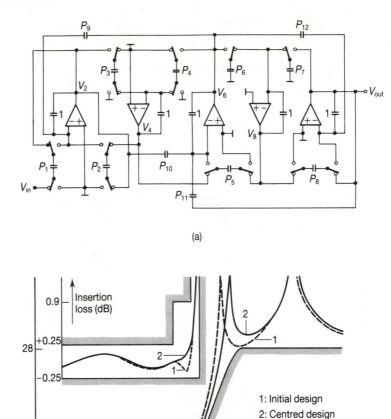

Figure 8.13

(a) Circuit diagram of switched-capacitor filter with 12 design parameters.   (b) Nominal response and performance specifications of switched capacitor filter example.

value of unity for X_0, associated with a predicted yield increase from 0.24 to 0.36, as offering a reasonable compromise between predicted yield increase and the confidence attached to that prediction. The capacitor values corresponding to this choice of X_0 were then computed, and the whole exercise repeated beginning with a fresh 500-sample Monte Carlo analysis. After four complete steps a design centre with an estimated yield of 0.54 was obtained. To check the effectiveness of both the yield and variance prediction formulae, separate Monte Carlo analyses were carried out using capacitor values corresponding to the different values of X_0 examined. The results shown in Figure 8.11 indicated, not unexpectedly, closer agreement for smaller values of X_0.

It should be noted that, as an optimum is approached, the yield prediction curve (e.g. Figure 8.11(a)) becomes very flat. Since only small improvements in yield are now attainable, the sample size of the Monte Carlo analysis needs to be increased in order to distinguish small differences in yield.

8.5   Conclusions

This chapter has discussed two Monte Carlo based methods of tolerance design that employ yield gradients. Of the two, parametric sampling has been developed for and applied to a variety of different tolerance design problems. The method employing yield prediction formulae, on the other hand, has primarily been developed for design centring, although its extension to tolerance assignment has been reported by Armaos (1981).

Both methods are difficult to implement when compared with the methods of the previous chapter. The yield prediction formulae developed by Antreich apply for component pdfs which are Gaussian. It would appear that entirely new formulae would be needed for different component pdfs.

8.6   Appendix

A detailed derivation of the yield and variance prediction formulae employed by Antreich and co-workers will be found in the reference (Antreich and Koblitz, 1982). In this appendix we outline some of the basic steps to illustrate the general principle.

Firstly for notational convenience we write yield as the expectation of the testing function $g(P)$ with respect to the pdf $\Phi(.)$. That is,

$$Y_1 \triangleq Y_1(P_1^0) = \int \ldots \int g(P)\Phi(P, P_1^0)\mathrm{d}P \tag{8.27}$$

is written as

$$Y_1 = E_1[g(P)] \tag{8.28}$$

where $E_1(x)$ implies the expectation (average) of x with respect to the pdf $\Phi(P, P_1^0)$.

In design centring the form of the component pdf is invariant and only changes in the design centres (component nominals) are being explored. The present exercise requires a formula for predicting yield associated with an exploratory design centre P^0, by extrapolation of the results of a conventional Monte Carlo analysis performed for a different design centre P_1^0.

In common with the parametric sampling estimators, the YPFs are based on the importance sampling relationship which allows an estimation of yield when the pdf from which the Monte Carlo samples are taken is different from the applicable component pdf. In this case the Monte Carlo samples are taken from the pdf $\Phi(P, P_1^0)$ whereas the relevant component pdf is $\Phi(P, P^0)$. The importance sampling relationship allows us to write the yield associated with $\Phi(P, P^0)$ as

$$Y(P^0) = \int \ldots \int g(P) \frac{\Phi(P, P^0)}{\Phi(P, P_1^0)} \Phi(P, P_1^0) \, dP \tag{8.29}$$

or more succinctly as

$$Y(P^0) = E_1 \left[g(P) \frac{\Phi(P, P^0)}{\Phi(P, P_1^0)} \right] \tag{8.30}$$

The YPFs are based on Expressions 8.29 and 8.30. To develop these further some assumption or knowledge of the particular form of the pdf $\Phi(.)$ is required. Following Antreich we take $\Phi(.)$ to be multivariate normal:

$$\Phi(P, P^0) = \frac{1}{\sqrt{(2\pi \det [C_0])}} \exp \left[(P - P^0)^T [C_0]^{-1} (P - P^0) \right] \tag{8.31}$$

where C_0 is the variance–covariance matrix of $\Phi(.)$. That is, C_0 is a $(k \times k)$ matrix where k is the number of component parameters and the element c_{ij} is the covariance between component parameters p_i and p_j. The diagonal elements such as C_{ii} represent the variance of the ith component p_i. For purposes of design centring the elements of the matrix C_0, which represent component spreads (tolerances) and correlations, remain invariant, but will appear in the YPF.

After substituting multinormal expressions such as Equation 8.31 for the functions $\Phi(P, P^0)$ and $\Phi(P, P_1^0)$ in Equation 8.30, and some algebraic manipulation, we get

$$Y(P^0) = E_1 \left[\frac{g(P) \sqrt{\det [C_0]} \exp \{ -\frac{1}{2}(P - P^0)^T C^{-1}(P - P^0) \}}{\sqrt{\det [C_1]} \exp \{ -\frac{1}{2}(P - P_1^0)^T C_1^{-1}(P - P_1^0) \}} \right] \tag{8.32}$$

However, for design centring, $C_1 = C_0 = C$ and therefore

$$Y(P^0) = E_1 [g(P) \exp \{ (P - P^0)^T C^{-1} \Delta P - \frac{1}{2} \Delta P^T C^{-1} \Delta P \}]$$

$$\Delta P \triangleq P^0 - P_1^0 \tag{8.33}$$

The next two steps are, first, to expand the terms in brackets via a Taylor series expansion truncated after the second-order terms and, second, to take the expectation with respect to the pdf $\Phi(P, P_1^0)$ for the individual terms. This leads to the expression

$$Y(P^0) \sim Y_1 \{1 + (P_H - P_1)^T C^{-1} \Delta P - \tfrac{1}{2} \Delta P^T C^{-1} (C - C_H)$$
$$- (P_H^0 - P_1^0)(P_H^0 - P_1^0)^T C^{-1} \Delta P)\} \qquad \textbf{(8.34)}$$

where two new terms P_H and C_H have been introduced. These are the means (P_H) and the variance–covariance matrix (C_H) of a new pdf, $h(.)$, obtained from $\Phi(P, P_1^0, C)$ by truncating with the testing function $g(P)$. That is:

$$h(P, P_H^0, C_H) = Y^{-1} g(P) \Phi(P, P_1, C) \qquad \textbf{(8.35)}$$

The terms P_H and C_H are given by:

$$P_H = E_1[Y_1^{-1} P g(P)]$$
$$C_H = E_1[Y_1^{-1}(P - P_H^0) g(P)] \qquad \textbf{(8.36)}$$

Expression 8.34 is a useful yield prediction formula. Careful examination will show that it allows us to predict the yield for a trial design centre P_1^0 and various parameters of the component pdfs.

CHAPTER 9

The Use of Sensitivity Analysis

OBJECTIVES

For many aspects of circuit performance it is now a simple matter to compute the effect on one or more properties of a circuit of small changes in all the component values. Such **component sensitivity** information is not only of immense benefit to the designer in its own right, but also has the potential for considerably reducing the computational effort involved in tolerance design. The reason is simple: if the performance of a circuit represented by a particular point in component space has been computed − at some cost − by a circuit analysis package, then a knowledge of component sensitivities allows the circuit's performance at a nearby point to be approximately determined at minimal cost. Since the nearby point could be a point selected by the Monte Carlo sampling process, the potential for reducing the cost of a Monte Carlo analysis is clear. In this chapter the application of sensitivity information is illustrated within the context of tolerance assignment.

9.1 Sensitivity analysis

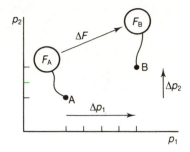

The problem for which inexpensive sensitivity analysis offers an attractive solution is illustrated in the (by now) familiar context of two-parameter space by Figure 9.1. The circuit has been analysed at point A and its performance F_A noted: bear in mind that a considerable sum may have been spent to determine F_A in this way. We now wish, for one of many reasons (some of which we mention shortly), to know the circuit performance F_B at a nearby point B. Movement from A to B involves a displacement Δp_1 along p_1 and a displacement Δp_2 along p_2. Using a truncated Taylor series approximation to the circuit performance F we can write the unknown F_B in terms of the known F_A.

$$F_B \cong F_A + \left. \frac{\partial F}{\partial p} \right|_A \Delta p^T \qquad (9.1)$$

Figure 9.1
The circuit performance associated with two nearby points in parameter space.

where, for the example of Figure 9.1,

$$\Delta p = [\Delta p_1 \Delta p_2]$$

The quantity $\partial F / \partial p$ is known as the differential sensitivity of F with respect to the parameters, and in the illustration of Figure 9.1 would be a two-element vector. Thus, for example, if

$$\frac{\partial F}{\partial p} = [2 \quad 1]$$

and if

$$\Delta p = [\Delta p_1, \Delta p_2] = [4 \quad 2]$$

then the approximate value of the change ΔF in F on moving from A to B would be

$$\Delta F = [2 \quad 1] \begin{bmatrix} 4 \\ 2 \end{bmatrix} = 10$$

Clearly, the calculation of the approximate value of F at other points in parameter space involves negligible cost once the differential sensitivity is known. The results are, of course, approximate, and one could attempt to improve the approximation by using more terms in the Taylor series: however, the calculation of second- and higher-order sensitivities that this would require is not a realistic proposition. In practice, one uses the approximation of Equation 9.1, and makes occasional checks by means of other calculations.

We have seen how knowledge of the differential sensitivity of circuit performance with respect to parameter change allows the performance at other points in parameter space to be determined approximately. The following section shows how differential sensitivity can be computed inexpensively.

9.2   The transpose circuit method

A full description of methods of sensitivity analysis appropriate to electronic circuits would be out of place here, and can in any case be obtained elsewhere (Brayton and Spence, 1980; Director and Rohrer, 1969). Here we shall briefly describe, without theoretical justification, how the differential sensitivity of the frequency-domain behaviour of a circuit, at a single frequency, can be computed.

Suppose that the circuit of interest is the linear model of an amplifier (Figure 9.2). It contains many components, but for clarity we only show three of them: a resistor of conductance G, a capacitor C and a voltage-controlled current source having a mutual conductance g. The latter component may, for example, be part of the hybrid-π model of one of the transistors within the amplifier. The circuit performance of interest is the 'output' voltage V_{out}.

It is a straightforward matter, by means of a circuit analysis package, to compute, in addition to the voltage V_{out}, the voltages across all internal components (Figure 9.3). Thus, the (probably complex) voltages across G and C are V_G and V_C respectively. That across the terminal pair whose voltage controls the current source is V_P and that across the controlled source V_Q.

Having noted the results of analysing the circuit of Figure 9.2, we now consider the analysis of a new circuit which is, nevertheless, closely related to the old one. It is, for example (Figure 9.4), topologically identical: where there is a two-terminal component in the old circuit, there is also a two-terminal component – of the same value – in the new

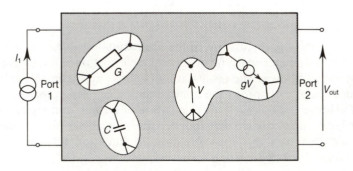

Figure 9.2

The linear model of an amplifier with components of interest exposed.

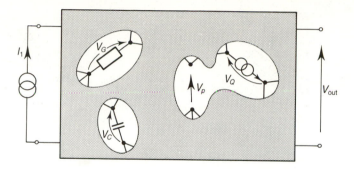

Figure 9.3
The voltages associated with some
internal components.

circuit. Where there is a controlled source in the old circuit there is still
a controlled source in the new one, but with the locations of the
controlling voltage and the controlled current interchanged; the
parameter value (here the mutual conductance) remains unchanged.
The new circuit of Figure 9.4 is called the **transpose circuit**. The
excitation of the transpose circuit also differs from that of the original
circuit. Where, before, there was a current excitation at port 1 we now
substitute an open circuit: where there was an 'output' voltage at port 2
we now connect a current excitation source (Figure 9.4) of 1 A.

The transpose circuit is now analysed by the same analysis
package as used for the original circuit, and at the same frequency.
Again, we note the voltage across each internal component. We denote
these voltages in the same way as before (Figure 9.5), but with a
superscript T to denote association with the transpose circuit. Thus, the
voltage across the capacitor is V_C^T.

The calculation of sensitivities is now a simple matter. It can be
shown (Brayton and Spence, 1980), for example, that the sensitivity of
V_{out} to small changes in the conductance G of the resistor is given by

$$\frac{\partial V_{out}}{\partial G} = -V_G V_G^T \qquad\qquad (9.2)$$

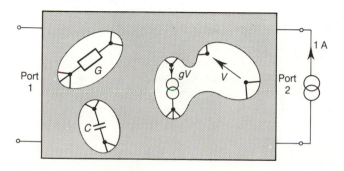

Figure 9.4
The transpose circuit.

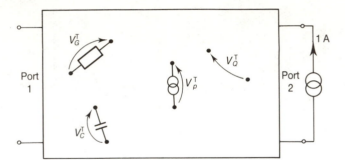

Figure 9.5

Response of the transpose circuit.

In other words, the calculation of sensitivity involves only the product of two voltages. The same relation applies to the sensitivity of V_{out} to the susceptance ($j\omega C$) of the capacitor. For the voltage-controlled current source the sensitivity to small changes in the mutual conductance g is given by

$$\frac{\partial V_{out}}{\partial g} = - V_P V_Q^T \tag{9.3}$$

which is an equally simple calculation. Note that all these sensitivities are those of V_{out} with respect to parameter values. To determine the sensitivity of any other voltage, the analysis of the transpose circuit would have to be repeated with the 1 A current source appropriately relocated.

The important conclusion we can draw from the above discussion is that the sensitivity of one voltage (e.g. V_{out}) with respect to small changes in *all* component parameters can be obtained at the cost of only two circuit analyses. Without this method, determination of the sensitivities would require a perturbation approach involving $N + 1$ circuit analyses (for an N-parameter circuit) in which the effect of parameter changes are analysed one at a time. In fact – though we shall not prove it here – it is possible, given suitable software, to arrange that the calculation of frequency-domain (and d.c.) sensitivities involves only *one* circuit analysis and negligible additional calculation (Laksberg, 1978). In this case, sensitivity information is virtually free. Unfortunately, for time-domain performance, the procedures are not so straightforward: the interested reader is referred to Brayton and Spence (1980).

Having established that sensitivity information can be obtained economically, the remainder of this chapter will illustrate how this fortunate situation can be exploited to the benefit of tolerance design. The example selected is that of tolerance assignment.

9.3   Tolerance space

In view of its convenience and generalizability we again consider the simple case of a circuit (Figure 9.6) containing just two toleranced components. Since it is usual for design centring to be carried out prior to tolerance assignment and since, also, the design centre is probably essentially independent of component tolerances, we shall assume that the nominal values of the parameters p_1 and p_2 are fixed. Under this assumption the tolerance region R_T (Figure 9.7(a)) defined by the tolerances on p_1 and p_2 is equivalently defined by a single point in (for this example) two-dimensional space having the parameter tolerances as axes (Figure 9.7(b)). This latter representation – of R_T in tolerance space – is central to the discussion which follows.

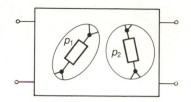

Figure 9.6
A two-parameter circuit.

9.4   Unit circuit cost

The aim of tolerance assignment is to minimize the cost of the manufactured circuits that meet the customer's specification. It is therefore necessary, at the outset, to define a satisfactory cost model. There are many such models, for most of which the tolerance assignment algorithm to be described would be satisfactory. The one we shall use for illustration assumes, first, that the cost (in dollars, pounds or whatever) of a component is equal to a constant plus a contribution which is inversely proportional to its tolerance. Thus, the total cost of each M-component circuit (whether or not it passes the specification) can be expressed as

$$b + \sum_{i=1}^{M} \frac{a_i}{t_i} \tag{9.4}$$

Figure 9.7
The representation of a tolerance region of fixed nominal value but variable tolerances by a point in tolerance space.

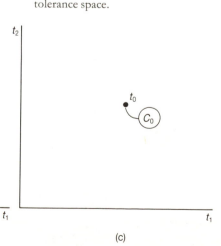

where t_i is the tolerance of the ith component, and a_i and b are constants. The constant b, for example, might represent labour costs, factory overheads and similar factors. However, because this cost must be recouped by the sale of only the satisfactory circuits, which constitute a fraction Y of the total (where Y is the yield), the unit circuit cost C for each *satisfactory* circuit is

$$C = \frac{1}{Y} \left[b + \sum_{i=1}^{M} \frac{a_i}{t_i} \right] \tag{9.5}$$

It is the quantity C, therefore, that we must minimize. The designable variables are the tolerances t_i associated with the parameters.

We note from the expression for C that, in order to carry out the minimization, we must know, to a satisfactory approximation, the value of the manufacturing yield Y. This requirement is common to all cost models to be used in tolerance assignment. For example, the cost model

$$C = \sum_{i=1}^{M} \frac{a_i}{t_i} \qquad \text{subject to } Y \geqslant Y_{\min}$$

involves the yield as a constraint, so that at some stage Y must still be estimated within acceptable confidence limits.

9.5   Tolerance perturbation

If our objective is to minimize the unit circuit cost C we must obviously obtain some estimate of C for the initial choice of component tolerances (Figure 9.7). This estimate is found by carrying out a Monte Carlo analysis within the initial tolerance region R_T, and substituting the resulting estimate $\hat{Y}$ of the yield for Y in Equation 9.5. Typically, 50 to

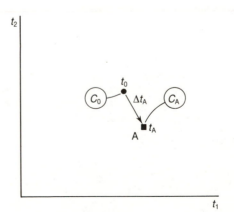

Figure 9.8
A new choice (A) of tolerances.

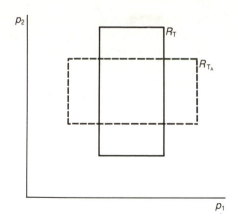

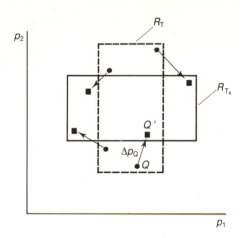

Figure 9.9
The new tolerance region R_{T_A}
associated with the tolerances
represented by point A in
Figure 9.8.

Figure 9.10
The original and new tolerance
regions, with related samples.

100 samples might be employed to obtain a reasonably accurate estimate of both yield and cost. Thus, for one point (shown as t_0 in Figure 9.7(b)) in tolerance space we have an estimate of the cost C_0 (Figure 9.7(c)).

With a view to reducing C we now explore the effect of different choices of parameter tolerances: i.e. different choices of t_1 and t_2. Such choices are usefully represented first in tolerance space (Figure 9.8). The initial choice of tolerances was represented by a single point (t_0) with which a cost C_0 was associated. A new choice is represented by the point A, and we would obviously like to know the cost (C_A) associated with this choice (t_A) of parameter tolerances.

The point A in tolerance space defines a new tolerance region R_{T_A} in parameter space (Figure 9.9). To find the associated cost C_A we clearly need, according to Equation 9.5, an estimate of the yield associated with R_{T_A}. We could, of course, carry out a fresh Monte Carlo analysis, but this would be expensive if, as we intend, we explore the costs associated with a number of points in tolerance space, i.e. a number of tolerance regions. An alternative, and far less expensive, approach involves the use of sensitivities.

9.6 Use of sensitivities

Figure 9.10 again shows the original tolerance region R_T and the new tolerance region R_{T_A}. Whereas, during the Monte Carlo analysis associated with R_T, around 50 or 100 samples may have been used, we choose for clarity to show the location of just four of them. To show that they are associated with R_T, we denote them by circular blobs to conform to the representation of Figure 9.8. Now assume that the four corresponding Monte Carlo samples to be associated with the new tolerance region R_{T_A} are distributed geographically within that region in the same way as for R_T. In other words, both R_T and R_{T_A} possess the same relative spatial location of samples. We can imagine that R_T is painted on a membrane that is stretched/compressed to produce R_{T_A}, and that the sample points have been correspondingly translated. The four sample points within R_{T_A} are indicated by square blobs in Figure 9.10. Such a selection of sample points within R_{T_A} seems not unreasonable if the yields, and hence costs, associated with R_T and R_{T_A} are to be compared.

The translations involved from the points within R_T to the corresponding points within R_{T_A} have been indicated by vectors in Figure 9.10. We now decide that, rather than carry out fresh circuit analyses at the new sample points (four of which are shown) we instead calculate them approximately by use of the Taylor series expression of Equation 9.1. Thus, with reference to the two sample points Q and Q′, we would assume that in the original Monte Carlo analysis both F and $\partial F/\partial p$ have been computed for point Q, whereupon $F_{Q'}$ associated with point Q′ is computed according to

$$F_{Q'} = F_Q + \left.\frac{\partial F}{\partial p}\right|_A \Delta p_Q$$

where Δp_Q is the displacement of Q′ from Q. Values of F at Q′ and other sample points within R_{T_A} would then be checked against specifications, and the estimate $\hat{Y}_A$ of yield obtained as with any normal Monte Carlo analysis. Substitution in Equation 9.5 then provides an estimate of the cost C_A associated with the point A in tolerance space.

To summarize we see that, following a Monte Carlo analysis of the original tolerance region R_T, in which both circuit performance F and its differential sensitivity $\partial F/\partial p$ with respect to all toleranced parameters are computed for each sample, the estimation of unit circuit cost C associated with a new set of tolerances is both straightforward and inexpensive.

9.7   Search of tolerance space

As we have just seen, the computational burden of estimating the unit circuit cost C for a new set of parameter tolerances (represented by a single point t in tolerance space) is sufficiently small to allow a

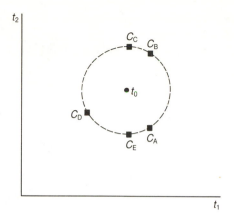

Figure 9.11
The random selection of tolerances on a circle having centre t_0. Each point is associated with a circuit cost.

generous sampling of tolerance space. The results are only approximate, of course, but, provided the tolerance perturbation Δt is not large and suitable checks are made during the tolerance assignment algorithm, the results are acceptable.

Quite empirically we decide to select tolerance perturbations Δt_i ($= t_i - t_0$) randomly on a hypersphere in tolerance space having as its centre the original set of tolerances t_0. In the case of our two-dimensional example this means the random selection of points on a circle with centre t_0. Five such points are shown in the illustration of Figure 9.11. Mindful of the error inherent in the Taylor series approximation of Equation 9.1 we might prudently limit the radius of the circle to one-tenth of the magnitude of t_0.

The result of such an exploration of tolerance space will be a set of points with associated estimated costs, as shown in Figure 9.11. The information obtained by this exploration of tolerance space must now be used to guide the choice of a direction, in that space, that is most likely to lead to a minimum of C.

9.8   Search direction

As described above, the sampling of tolerance space has produced a result, such as that shown in Figure 9.12, where the numerical value of the cost associated with each of a number of points in that space has been estimated. How, by examination of these costs ($\hat{C}_0$, $\hat{C}_A$, $\hat{C}_B$, $\hat{C}_C$, $\hat{C}_D$ and $\hat{C}_E$ in Figure 9.12(a)), can we determine the direction of movement in tolerance space likely to lead to a minimum of C?

There is no unique way of determining such a search direction. One, leading to a 'weighted average' search direction, involves the

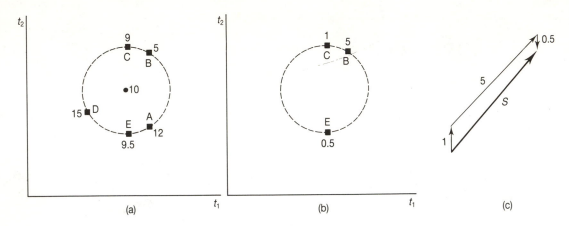

Figure 9.12
Determination of the search
direction.

following steps. First, to each tolerance perturbation (i.e. for each point other than the original, t_0) is assigned a weight W which is simply the difference between the original cost C_0 and the cost associated with the perturbed tolerance: i.e., for the point A (Figure 9.12(a)),

$$W_A = \hat{C}_0 - \hat{C}_A = -2$$

Next, those perturbations characterized by a positive weight are identified (Figure 9.12(b)) and all others temporarily ignored. The weighted average search direction is then given by

$$S = \frac{1}{J} \sum_{i=1}^{J} W_i \Delta t_i \qquad (9.6)$$

where those perturbations having a positive weight are considered to be numbered consecutively from 1 to J. The neglect of negatively weighted points is supported by practical experience, which has shown that a better indication of the descent direction for C is provided by Equation 9.6, especially in the neighbourhood of a local optimum. The determination of search direction according to Equation 9.6 is illustrated by Figure 9.12(c) for the numerical example of Figure 9.12(b).

9.9   Choice of step length

Once the search direction (S) has been chosen, a step of length d in that direction will define a new set of parameter tolerances according to

$$t_{0,\text{new}} = t_0 + S \cdot d$$

Normally, a search will take place in the direction S to determine that value of d associated with a minimum cost along that direction. This may not be – and usually is not – the absolute minimum of C, but simply the lowest value that can be achieved if movement is constrained to the search direction. Thus, with the cost model of Equation 9.5, we are faced with a one-dimensional minimization problem

$$\underset{d}{\text{Minimize}} \ C(t_0 + S \cdot d)$$

In other words, along the direction S we must minimize the cost C (which is a function of the original tolerances t_0 and the distance d moved along S) by a suitable choice of the step-length d.

Fortunately for us, little expense is involved in computing the value of C for a number of points in tolerance space along the search direction S. The reason is that we can again make use of the sensitivity information computed during the original Monte Carlo analysis. Thus, for a given choice of d, corresponding to a new point in tolerance space and hence a new tolerance region in parameter space, the same relative spatial location of samples in the tolerance region is chosen as for the initial tolerance region, and the corresponding values of F calculated according to (9.1). From the computed values of F the yield can be estimated and hence, according to the cost model (Equation 9.5), the value of C. If a uniform line search is carried out along S a result such as that shown in Figure 9.13 may be obtained. The search would normally continue until (as in Figure 9.13) a minimum cost point is

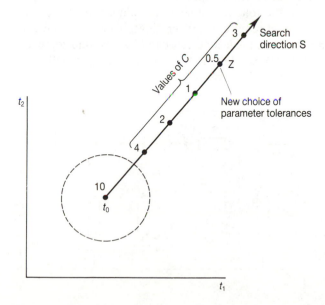

Figure 9.13
Search to find a minimum of the cost function C along the search direction S.

bounded. Then, the point associated with the minimum cost (point Z in Figure 9.13) would be chosen as defining the new initial set of tolerances, and the entire procedure repeated.

9.10   Termination criterion

Above we have described one complete step in the reduction of C. This step comprises the following:

(1)   A Monte Carlo analysis is carried out for the initial tolerance region, but one in which performance sensitivity is computed for each sample. From the results, both Y and C are estimated.

(2)   Tolerance space is sampled: for each point in tolerance space the yield is computed using sensitivity information, and the cost C determined.

(3)   A search direction is found from those samples in tolerance space which exhibit a decrease in cost from the initial choice of tolerances.

(4)   A uniform search along the search direction is carried out to find a minimum of C.

(5)   This minimum is chosen as defining a new, improved set of tolerances.

In the course of tolerance assignment the above procedure will normally be repeated until further application is judged not to be worthwhile. In other words, the tolerance assignment algorithm is typically iterative. Being iterative it requires some termination criterion. One possibility is to terminate when the last decrease in C is less than some fixed fraction of the current cost. Often, the circuit designer will exercise some judgement concerning termination.

Figure 9.14
The circuit for which tolerance assignment is to be carried out.

Component	Nominal value
C_1	11.93 nF
C_2	8.85 nF
C_3	11.13 nF
C_4	37.40 nF
C_5	94.67 nF
L_1	3.90 mH
L_2	3.09 mH

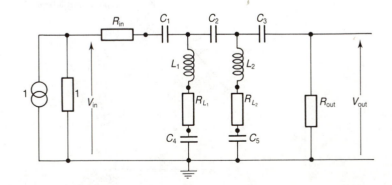

9.11   Practical examples

For the high-pass filter of Figure 9.14 the performance of interest is the attenuation relative to that at the reference frequency of 990 Hz. The nominal performance, together with performance specifications, is shown in Figure 9.15. A sample circuit is deemed to be acceptable if the performance at 12 selected frequencies (170, 350, 440, 630, 650, 720, 740, 760, 940, 990, 1040 and 1800 Hz) satisfies the specifications. The circuit contains seven independent toleranced parameters, each of which has a truncated normal distribution: the nominal values correspond to the mean, and the tolerances to the 3σ truncation limits. Two resistive parameters, introduced to model the loss in the inductors, are completely correlated with the corresponding inductance values. The cost model chosen for the filter is

$$C = \frac{1}{Y} \left\{ 2.5 + \sum_{\text{capacitors}} \frac{1}{t_i} + \sum_{\text{inductors}} \frac{2}{t_i} \right\}$$

which is of the same form as Equation 9.5.

The nominal values of the parameters were fixed at the values shown in Figure 9.14, and initial tolerances were set at 5% of nominal. Under these initial conditions the yield estimate was 100% and the cost C was estimated to be 4.3. The tolerance assignment algorithm described in this chapter was then applied.

At each iteration 50 Monte Carlo samples were analysed. Then, 25 tolerance perturbations were used to explore tolerance space. Following this a search direction was determined though using a

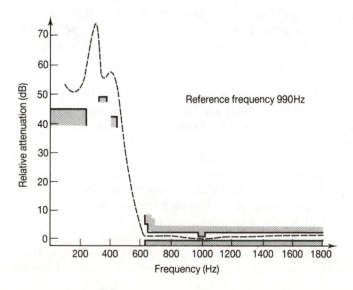

Figure 9.15
Nominal response and specifications of high-pass filter.

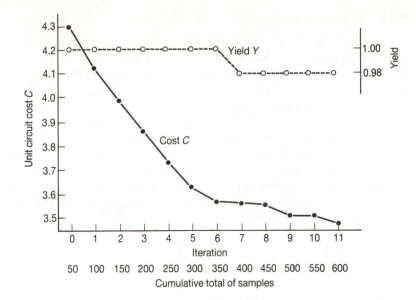

Figure 9.16
Variation in cost and yield during
tolerance assignment.

slightly different method to the one described above (Ilumoka and
Spence, 1982). A one-dimensional search, as described in Section 9.9,
then indicated the new set of parameter tolerances.

The resulting variation in unit circuit cost is shown in Figure 9.16,
and the variation in component tolerances in Figure 9.17. There is of
course no way of knowing if the result really does correspond, within an
acceptable margin, to the actual optimum design. It should therefore be
mentioned that, for the same tolerance assignment task, other workers
(e.g. Singhal and Pinel, 1981) have obtained essentially identical results.
Indeed, the reader should be aware that, during the development of
tolerance design methods, different researchers wisely selected common
problems in order to enhance both confidence and comparison.

An understandable reaction to the above example concerns the
number of circuit analyses involved, a total of 600. While the benefit of
the tolerance assignment exercise may be so high that the computational cost is more than justified, the algorithm would be even more
attractive if fewer analyses were involved. Currently, we just do not
know how few analyses will suffice, but a relevant and interesting
experiment has thrown light on this question. For the same tolerance
assignment task described above it was decided to use very few Monte
Carlo samples at the outset, and then increase the number of samples as
the algorithm proceeded. Additionally, it was acknowledged that the
human designer's own judgement might enhance the procedure, so the
user was allowed to monitor progress and choose the number of Monte
Carlo analyses for any iteration. The result is shown in Figure 9.18.

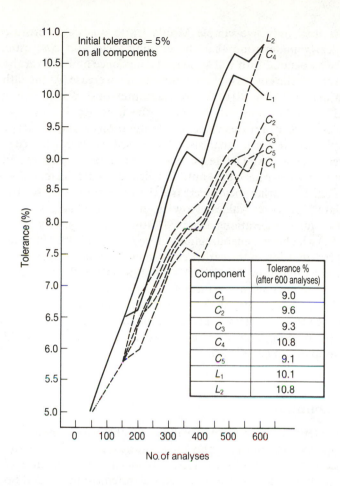

Figure 9.17
Component tolerance variation
during tolerance assignment on
high-pass filter.

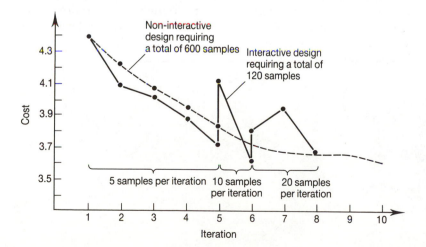

Figure 9.18
Interactive design of high-pass filter
using small sample sizes.

At first, only five-sample Monte Carlo analyses were used, it being clearly understood that the ensuing yield and cost estimates would be associated with widely spaced confidence limits. Over the first four iterations the cost exhibited a monotonic decrease. At the fifth, the yield estimate fell from 1.0 to 0.8 because a previously acceptable circuit failed the specifications. At this point the user decided to reset the tolerances to their previous value (i.e. for iteration 4), and to carry out a 10-sample Monte Carlo analysis. The revised estimates for cost and yield were then 4.02 and 0.9 respectively. The same sample size (10) was used for two further iterations until the yield estimate decreased whereupon, as at the earlier yield decrease, the tolerances were set to their immediately previous values. Now, using an increased sample size of 20, two further iterations achieved estimated cost and yield values (3.58, 0.95) which were encouragingly close to those (3.48, 0.98) achieved at the 11th iteration in the earlier experiment in which 50 samples were used at each iteration. Moreover, a total of only 120 circuit analyses were employed in contrast to the 600 used earlier.

While the results of the experiment should be treated with caution, it serves to encourage further study both of the benefits of interactive working and of the extent to which the number of Monte Carlo samples can be reduced.

9.12   Comment

This chapter's main aim was to illustrate the part that inexpensive sensitivity information can play in tolerance design. The activity of tolerance assignment was used to provide a demonstration of this potential, though it has also been demonstrated (Ilumoka and Spence, 1982) in the context of variability reduction as well as design centring. For the case of tolerance assignment only outline considerations have been presented here, and the reader who intends to implement the algorithm would be well advised to consult a more detailed exposition (Ilumoka and Spence, 1982).

CHAPTER 10

Questions and Answers

OBJECTIVES

In the interest of brevity, we have had to be extremely selective about our choice of material in the foregoing chapters. We attempt to remedy a number of important omissions by including in the present chapter answers to questions frequently raised concerning tolerance design. Indeed, almost all the questions discussed have been asked (mainly by industrial delegates) during one or more of the course presentations on which this book is based. The other objective of the chapter is, via the answers to selected questions, to point the reader who wishes to dig deeper into the subject towards key papers and significant developments.

Question 1   **Comparative performance of tolerance design methods**

Has any comparative evaluation of tolerance design methods been carried out in order to guide anyone who has to make a choice of which algorithm to implement?

Answer 1   Yes. In a conference paper, Wehrhahn and Spence (1984) presented the results of applying eight tolerance design methods to nine tolerance design problems. The problems ranged widely in the number of adjustable parameters, from two to 43, and the application of the tolerance design method was carried out by the creator of the algorithm or his designate. There was never any intention of identifying the 'best' algorithm, since each has its own characteristics and it would be impossible to propose a single figure of merit. The number of circuit analyses required to achieve a solution (not including the Monte Carlo analysis required to confirm that solution) ranged from below 10 in certain cases to almost 10 000. The computational effort ranged from seconds to hours, though about 5 s CPU on a VAX/780 was not unusual. The computational effort depends upon the analysis method, the size and type of circuit, the number and type of design constraints and the goodness of the starting point.

Question 2   **Tolerance design methods suited to integrated circuits**

Most of the tolerance design methods described are suited to discrete circuits, and have principally been illustrated in that context. Are new methods being developed specifically for integrated circuits, where the number of designable parameters for a given chip may number in the hundreds or thousands?

Answer 2   They are. Perhaps a first reference to consult in this connection is the January 1986 issue of *IEEE Transactions on Computer-Aided Design of Integrated Circuits and Systems*. It's a special issue on the statistical design of VLSI circuits, and contains 12 papers on this topic. A brief summary of the content of one paper may help to answer the question.

Dr. Ping Yang and his associates (Yang *et al.*, 1986) concluded, from experimental data and other considerations, that only *four* process parameters are responsible for the variation of the behaviour of small-dimension NMOS logic circuits. Of these four parameters two are geometrical (length reduction and width reduction), a third is oxide thickness and the fourth is the flat-band voltage. As a consequence, and under the reasonable assumption (for the conditions considered) of linearity, only five circuit simulations are necessary in order to determine the coefficients relating circuit performance variation to the four parameters. Once these coefficients are known, the linear equations that describe the boundaries of the region of acceptability in four-dimensional process parameter space are obtained, this region being called the 'yield body'. The yield can then be predicted by integrating the pdf over this yield body. The integration is carried out by the random generation of points within the tolerance region, each point being tested to see if it lies inside the yield body: no additional circuit analyses are involved.

Yang then proceeds to describe a scheme for yield optimization which is illustrated by its application to a 10-transistor CMOS logic circuit. In four iterations the yield was increased from essentially zero to about 95%.

Question 3   **The existence of local and global optima**

When applying a yield maximization algorithm to a circuit, is there a danger of finding a local rather than the global optimum?

Answer 3   There is, and one example of this can be seen in the results reported for Problem 1 in the paper by Wehrhahn and Spence (1984). For this problem, the methods of Wehrhahn (1984), Antreich and Koblitz (1982) and Kjellstrom and Taxen (1981) appeared to reach a global optimum, while others achieved a local optimum.

Question 4   **Sensitivity methods applied to worst-case design**

Can sensitivities of the type discussed in Chapter 9 be applied to the task of worst-case design and design centring?

Answer 4   Yes. The concept of 'margin sensitivity' was devised by David Agnew (1980) and has since been incorporated in a package called SCAMPER. Briefly, as illustrated in Figure 10.1, one identifies those frequencies for which the margin between the actual performance and the

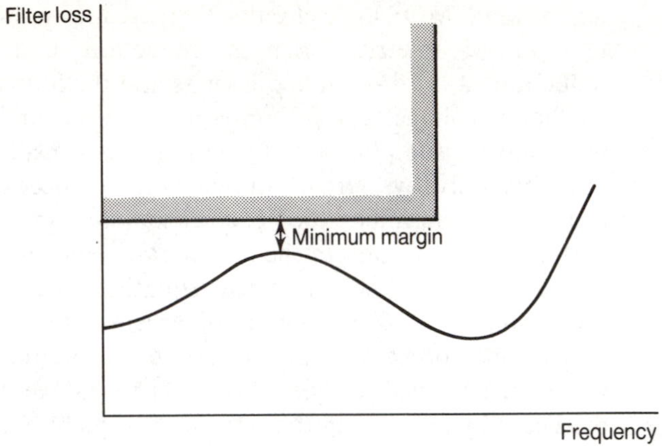

Filter loss

Minimum margin

Frequency

Figure 10.1
The minimum margin between
nominal performance and the
specification is identified with a view
to its maximization.

specification is at a minimum. Then, based on knowledge
of the sensitivity of circuit performance to parameter
change at those critical frequencies, parameter values are
adjusted in order to maximize the minimum value of the
margins. Maratos (1988) also devised a method of yield
optimization that exploits knowledge of sensitivities. As
a contribution to the comparison of methods carried out
by Wehrhahn and Spence (1984), Maratos' method was
applied to two problems and exhibited an impressive
economy of circuit analysis.

Question 5   **Statistical analysis of large circuits**

*If a circuit is very large, the cost of simulating its
performance by means of a package such as SPICE is
considerable. A Monte Carlo analysis involving (say) 200
circuit analyses would therefore be extremely expensive. Is
there any way of overcoming this disadvantage of a Monte
Carlo analysis?*

Answer 5   One solution to this problem has been provided by Soin
and Rankin (1985) through a concept known as the
control variate. The solution can perhaps best be
understood by reference to Figure 10.2. In addition to the
(large) circuit C_E under consideration, a companion or
shadow model C_S is required whose component
parameters are dependent upon those of the large circuit
and whose response approximates to that of C_E, but
which is much simpler and hence computationally
inexpensive to analyse. The basic procedure, illustrated in

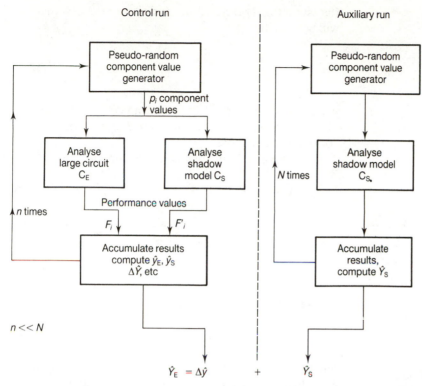

Figure 10.2
Illustrating the control variate procedure for efficient yield estimation.

Figure 10.2, involves two Monte Carlo analyses. In the first, referred to as the **control run**, a relatively small sample size (100 or so) is used and both the circuits are analysed using the same sets of random component values. The control run provides an estimate of the difference between the yield of the two circuits, and the accuracy of this estimation has been enhanced by the use of common random values. The procedure is that of correlated sampling discussed in Chapter 7.

A second Monte Carlo analysis (the 'auxiliary run') is now performed, in which only the shadow model is analysed. Since the cost of analysis of the shadow model is small, we may be able to afford to analyse a large number (say 10 000) and hence obtain a very accurate estimate of its yield. Finally, this accurate shadow model yield estimate is added to the estimate of yield differences to obtain an enhanced estimate of the yield of C_E.

For a conventional Monte Carlo analysis the computational cost is proportional to the sample size which, in turn, is inversely proportional to the variance of the estimator. Therefore, the efficiency of the new method will be given by

$$\eta = \frac{\sigma^2 T_E}{\sigma_E^2 T} \tag{10.1}$$

where σ_E^2 and T_E are the variance and computational cost of the new procedure and σ^2 and T correspond to the conventional method.

The control run* provides the estimate

$$\widehat{\Delta y} = \hat{y}_E - \hat{y}_S \tag{10.2}$$

and the auxiliary run† provides the estimate $\hat{Y}_S$. The enhanced estimate is then

$$\hat{Y}_E = \hat{Y}_S + \widehat{\Delta y} \tag{10.3}$$

The variance of $\hat{Y}_E$ is given by

$$\text{var}(\hat{Y}_E) = \text{var}(\hat{Y}_S) + \text{var}(\widehat{\Delta y}) \tag{10.4}$$

Since a large sample size was used to estimate Y_S we may, as a first approximation, ignore $\text{var}(\hat{Y}_S)$ so that, from Equation 10.4,

$$\text{var}(\hat{Y}_E) = \text{var}(\widehat{\Delta y})$$
$$= \text{var}(\hat{y}_E) + \text{var}(\hat{y}_S) - 2\,\text{cov}(\hat{y}_E, \hat{y}_S) \tag{10.5}$$

This brings us to the heart of the matter. A conventional Monte Carlo analysis would have provided an estimate $\hat{y}_E$ with variance $\text{var}(\hat{y}_E)$. There are now additional terms, namely $\text{var}(\hat{y}_S)$ and $2\,\text{cov}(\hat{y}_E, \hat{y}_S)$. However, if the covariance term is positive and of sufficient magnitude, the overall variance is reduced, resulting in a more efficient method than the conventional Monte Carlo analysis. In the method being described a positive correlation is achieved by suitably constructing the shadow model. In essence, the designer is trading off knowledge of the behaviour of the main circuit (which behaviour is referenced to construct the shadow model), against statistical uncertainty.

* Yields associated with the control run are denoted by lower-case y.
† Yields associated with the auxiliary and composite runs are denoted by upper-case Y.

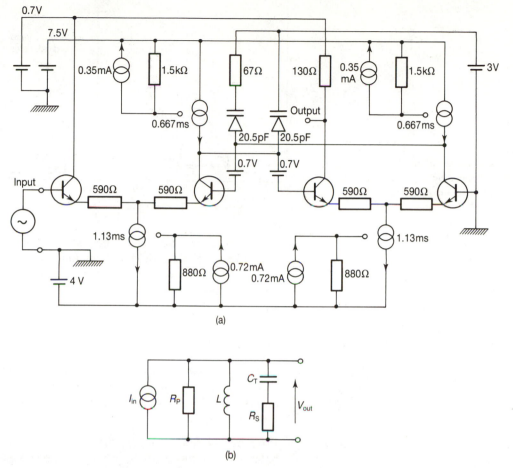

(a)

(b)

Figure 10.3

(a) Bipolar integrated band-pass filter: a large circuit. (b) Discrete counterpart of bipolar integrated circuit: the Shadow model. The components R_P, L, R_S, C_T, etc. are functions of the components of the large circuit above.

An illustration is provided by Figure 10.3, which shows the circuit under evaluation (a bipolar integrated band-pass filter) and its shadow model comprising a discrete component circuit. In this case the integrated circuit was intended as a replacement for the bulkier discrete circuit, and so the choice of shadow model was obvious. The histogram of Figure 10.4 shows, first, the results of 200 estimations of yield via the conventional Monte Carlo method, each estimate involving a sample size of 100. It also shows histograms for 200 yield estimates obtained by the control variate method. Whereas the average yield estimate in both cases is the same, the spread in the estimates in the case of the control variate method is smaller. Empirical studies for this example have yielded efficiencies ranging from about 3 to

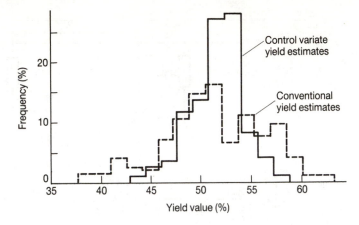

Figure 10.4
Yield estimates of the circuit of
Figure 10.3 obtained by
conventional and control variate
methods.

8. The method may easily be extended to estimated quantities other than yield: they include means, variances and other moments of performances and, additionally, yield sensitivities. Very much higher efficiencies have been reported for the estimation of these other parameters.

Question 6   **Black holes**

You have mentioned the concept of 'black holes', and shown that they can present difficulties in the tolerance design process. How likely are we to encounter black holes? Are they just a curiosity?

Answer 6   Probably the first person to draw our attention to black holes was Styblinski (1979), who presented some sections of parameter space relating to a filter circuit in order to display the black hole corresponding to the combination of that circuit and its specifications. It is generally believed that there is a far greater likelihood of encountering a black hole with a filter circuit having quite complex specifications than with, say, a logic circuit such as a NAND gate. One limitation to our knowledge of black holes is the enormous computational effort required to exhaustively search the parameter space of a realistic circuit.

Question 7   **Reduction of the cost of tolerance analysis by means of sensitivity information**

You mentioned in Chapter 9 that sensitivity analysis is inexpensive and, if used to estimate the circuit performance at Monte Carlo points, can thereby reduce the cost of tolerance analysis. Can you provide some feeling for the savings that can be achieved, and any advice that would be useful in realizing these savings?

Answer 7   Savings of about 70% of the required number of samples have been obtained with a Sallen-Key filter having seven toleranced parameters (Jones, 1986). When three methods of yield estimation (Jones, 1986) based on sensitivity information were applied to three different problems, savings ranged from about 40% to about 75%. These methods are heuristic, but were rigorously tested (Jones and Spence, 1987) by means of a large number of Monte Carlo analyses. Finally, remember that the sensitivity information may only be inexpensive if the number of circuit performances of interest is small, since a new analysis of the transpose circuit is required to compute the sensitivity of a given performance to all components. This statement should be qualified, however, since, if *LU* factorization is employed in a.c. analysis, the analysis of the transpose circuit takes place at negligible cost.

Question 8   **Application of tolerance design algorithms in other fields**

To what extent can the algorithms described in this book be applied outside the field of electronic circuits?

Answer 8   For systems such as space structures and mechanical mechanisms there is no reason why the algorithms should not be applied, since it is possible to simulate such systems on a computer: remember that all the tolerance design algorithms described require the simulation of samples of the circuit being designed. The picture is different, however, if we look at the manufacture of such artefacts as biscuits and ceramic tiles, and substances such as caramel. They have much in common with circuits, since for each of these artefacts or substances there are specifications: taste and robustness, perhaps, for biscuits, both of which are somewhat subjectively tested; dimensional reproducability for ceramic tiles; and viscosity for caramel. There are tolerances on parameters: the quality of flour supplied to the biscuit manufacturer, for example. And there are clearly designable variables, such as the lime content in the ceramic tiles (a crucial factor for dimensional reproducibility, it turns out). The significant difference, however, between electronic circuits on the one hand and such items as biscuits, tiles and caramel on the other is our ability to simulate. Under these circumstances the tolerance design algorithms reported in this book cannot be applied directly in these other domains, though some of the concepts introduced may well be valuable. Those who are interested in tolerance design in general may be interested to follow the recent literature on the so-called Taguchi method (Kackar, 1986) which is gaining momentum at present.

Question 9 **Tolerance design of microwave circuits**

None of the examples shown in the book has been concerned with microwave circuits containing such components as strip-lines. Is there any reason why the tolerance design algorithms should not be applied in this domain?

Answer 9 None at all. One example of a microwave circuit to which tolerance design has been applied is shown in Figure 10.5 (Hewitt, 1986, personal communication). The small-signal RF amplifier, of microstrip construction on a high dielectric substrate and with input and output matching circuits to 50 Ω impedance, was required to exhibit a gain between 10.8 dB and 12.8 dB over the frequency range from 0.8 to 1.2 GHz, to be unconditionally stable and to satisfy requirements regarding input and output VSWR. It was required to know the maximum allowable etch tolerance, the positional accuracy controlling line lengths, and the maximum tolerances of the lumped components.

The initial yield was estimated to be 20%. The

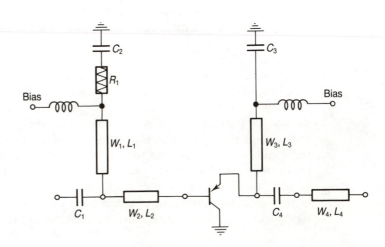

Figure 10.5
A microwave amplifier employing stripline. W = width, L = length.

tolerance design method used has much in common with the centres of gravity method, but with the interesting variant that the tolerance region that is explored expands and contracts (thereby requiring incremental sampling as in the common points scheme) as indicated by the direction defined by the CoG technique. The final estimated yield achieved was 98.7%.

Question 10 **Region of exploration**

Can, for example, the centres of gravity method of design centring be improved by generating the Monte Carlo points, not within the tolerance region, but within a somewhat larger region one might term the region of exploration?

Answer 10   Yes, though there is very little record of this approach except for its use by Singhal and Pinel (1981) and Wehrhahn (1987). This question does draw attention to the fact that we might be artificially constrained to performing the Monte Carlo analysis within R_T: during the initial stages of design centring we are interested, not primarily in the yield, but in locating R_A relative to R_T, and it is possible that an inflated region of exploration may help. Indeed, if the maximum possible yield is 100%, a situation in which a conventional design centring algorithm may be slow to converge, then design centring with a region of exploration somewhat larger than R_T (see Figure 10.6) may locate a 100% yield design quite efficiently. However, in expanding beyond R_T to a region of exploration R_E one has to keep dimensionality in mind: double the range of each parameter, and the volume of

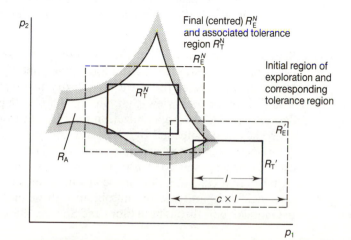

Figure 10.6

The use of a region of exploration, somewhat larger than the tolerance region, to assist the process of yield maximization.

the region to be searched is multiplied by 2^N, where N is the number of parameters. Recently, using a new and very simple algorithm, Wehrhahn (1987) successfully employed an inflation factor of between 1.2 and 2.

Question 11   **Zero yield**

If the yield of a circuit is estimated to be zero, what is the best course of action?

Answer 11   If the number of Monte Carlo samples employed in the tolerance analysis was large, then it is reasonably safe to assume that R_T lies wholly outside R_A. If this is so, then even the nominal circuit does not satisfy the specifications. The most appropriate step at this point might be to employ a conventional optimization package to adjust the *nominal* design to satisfy the specifications. An alternative would be to inflate the region of exploration beyond R_T, as discussed in Question 10 above, and carry out design centring if a pass circuit can be found. On occasion it may help to relax the specifications somewhat initially in order to find a pass point, and then restore them later when a reasonable yield has been achieved and the design process is progressing well.

Question 12   **Generality of the common points scheme**

Although the common points scheme described in Chapter 7 appears to be potentially valuable in reducing the cost of design centring by the centres of gravity method, it would seem to be applicable only to the case of uniform *component pdfs. Is this so, or can it be used if the parameter distributions are known, for example, to be Gaussian?*

Answer 12   Stein (1986) has shown that the common points scheme need not be restricted to uniform component pdfs. However, in view of the extra complexity involved, we would reiterate our view that, in the initial stages of design centring, yield maximization may benefit from, or at worst not be impeded by, the assumption of uniform component pdfs. An example of a design centring exercise which started with uniform pdfs and then switched to a Gaussian pdf for the last iteration showed a final yield comparable with that obtained through the use of Gaussian pdfs throughout (Soin and Spence, 1980).

References

Agnew, D. (1980) 'Design centering and tolerancing via margin sensitivity minimization.' *Proc. IEE, Pt.G*, **127**(6), 270–277

Antreich, K. J. and Koblitz, R. K. (1980) 'A new approach to design centering based on a multiparameter yield prediction formula.' *Proc. IEEE Int. Sym. Ccts. Sys. (ISCAS)*, Houston 1980, pp. 270–277

Antreich, K. J. and Koblitz, R. K. (1981) 'An interactive procedure for design centering.' *Proc. IEEE Int. Sym. Ccts. Sys.*, 1981, pp. 139–142

Antreich, K. J. and Koblitz, R. K. (1982) 'Design centering by yield prediction.' *IEEE Trans. Ccts. Sys.*, **CAS-29**(2), 88–96

Armaos, J. (1981) 'Ein allgemeines statistisches Verfahren zur Entwarfszentrierung und Toleranszuordnung.' *AEU*, **35**, 173–178

Bandler, J. W. (1974) 'Optimization of design tolerances using non linear programming.' *J. Optimization Theory Applications*, **14**, 99–114

Bandler, J. W., Liu, P. C. and Chen, J. H. K. (1975) 'Worst case network tolerance optimization.' *IEEE Trans.*, **MTT-23**(8), 630–641

Batalov, B. V., Belyakov, Y. N. and Kurmaev, F. A. (1978) 'Some methods for statistical optimization of integrated microcircuits with statistical relations among the parameters of the components.' *Soviet Microelectronics (USA)*, **7**(14), 228–238

Becker, P. W. (1974) 'Finding the better of two similar designs by Monte Carlo techniques.' *IEEE Trans. Reliability*, **R-23**(4), 242–246

Becker, P. W. and Jensen, F. (1977) *Design of systems and circuits for maximum reliability or maximum production yield*. McGraw-Hill, New York

Bjorke, O. (1977) *Computer aided tolerancing*. Tapir Publishers, Trondheim, Norway

Brayton, R. K. and Spence, R. (1980) *Sensitivity and optimization*. Elsevier, New York

Brayton, R. K., Hachtel, G. D. and Sangiovanni-Vincentelli, A. L. (1981) 'A survey of optimization techniques for integrated circuit design.' *Proc. IEEE*, **69**(10), 1334–1363

Director, S. W. and Hachtel, G. D. (1977) 'The simplicial approximation approach to design centering.' *IEEE Trans. Ccts. Sys.*, **CAS-24**(7), 363–372

Director, S. W. and Rohrer, R. A. (1969) 'The generalized adjoint network and network sensitivities.' *IEEE Trans. Cct. Theory*, **CT-16**, 300–336

Director, S. W. and Vidigal, L. M. (1981) 'Statistical circuit design: a somewhat biased survey.' *Proc. Eur. Conf. Cct. Theory Design* (*ECCTD*), 1981, The Hague, The Netherlands, pp. 15–24

Director, S. W., Hachtel, G. D. and Vidigal, L. M. (1978) 'Computationally efficient yield estimation procedures based on simplicial approximation.' *IEEE Trans. Ccts. Sys.*, **CAS-25**(3), 121–130

Elias, N. J. (1975) 'New statistical methods for assigning device tolerances.' *Proc. IEEE Int. Symp. Ccts. Sys.*, 1975, Newton, Mass., USA, pp. 329–332

Elias, N. J. (1979) 'The application of statistical simulation to automated analogue test development.' *IEEE Trans. Ccts. Sys.*, **CAS-26**(7), 513–517

Fletcher, R. and Powell, N. (1963) 'A rapidly convergent descent method for minimization.' *Computer J.*, **6**, 163–168

Godwin, J. J. (1955) 'On generalizations of Chebychev's inequality.' *J. Am. Stat. Assoc.*, **50**, 923–945

Hammersly, J. M. and Handscombe, D. C. (1964) *Monte Carlo methods*. Methuen, London

Hocevar, D. E., Lightner, M. R. and Trick, T. N. (1983) 'A study of variance reduction techniques for estimating circuit yield.' *IEEE Trans. Computer-Aided Design*, **CAD-3**(3), 180–192

Hocevar, D. E., Lightner, M. R. and Trick, T. N. (1984) 'An extrapolated yield approximation technique for use in yield maximization.' *IEEE Trans. Computer-Aided Design*, **CAD-3**(3), 279–287

Hooke, R. and Jeeves, T. A. (1961) 'Direct search solutions of numerical and statistical problems.' *J. Assoc. Computing Machinery*, **8**, 212–229

Ibbotson, I. R., Compton, E. and Boardman, D. (1984) 'Improved statistical design centering for electrical networks.' *Electronics Letters*, **20**(19), 757–758

Ilumoka, A. I. and Spence, R. (1980) 'Statistical approach to reduction of circuit variability.' *Electronics Letters*, **16**(20), 761–762

Ilumoka, A. I. and Spence, R. (1982) 'A sensitivity based approach to tolerance assignment.' *Proc. IEE, Pt.G*, **129**(4), 139–149

Ilumoka, A. I., Maratos, N. and Spence, R. (1982) 'Variability reduction: statistically based algorithms for reduction of performance variability of electrical circuits.' *Proc. IEE, Pt.G*, **129**(4), 169–180

Inohira, S., Shinmi, T., Nagata, M., Toyabe, T. and Iida, K. (1985) 'A statistical model including parameter matching for analog integrated circuits simulation.' *IEEE Trans. Computer-Aided Design*, **CAD-4**(4), 621–628

Ito, A., Gast, L. K., Coston, W. T., Lowther, R. E., Webb, R. W. and George, E. W. (1983) 'A statistical process and device simulator (SPADS).' *Proc. IEEE Int. Conf. Computer-Aided Design*, September 1983

Jones, I. W. (1986) *Performance optimisation of integrated circuit cells, with constrained parametric yield and process variation modelling,* PhD thesis, London University

Jones, I. W. and Spence, R. (1984) 'The optimisation of integrated circuit cells subject to a constraint on parametric yield.' *Colloquium on Design Software,* January 1984, Digest No. 1984/8, IEE, London

Jones, I. W. and Spence, R. (1987) 'Cost reduction of Monte Carlo yield estimates.' *Proc. IEE, Pt. G,* **134**(6), 249–258

Kackar, R. N. (1986) 'Taguchi's quality philosophy: analysis and commentary.' *Quality Progress,* December 1986, 21–29

Karafin, B. J. (1970) 'The optimum assignment of component tolerances for electrical networks.' *Bell Sys. Tech. J.,* **50**, 1225–1242

Kjellstrom, G. and Taxen, L. (1981) 'Stochastic optimisation in system design.' *IEEE Trans. Ccts. Sys.,* **CAS-28**, 702–715

Knauer, K. and Pfleiderer, H. J. (1982) 'Yield enhancement realized for analogue integrated filters by design techniques.' *Proc. IEE, Pt.G,* No. 4, 122–126

Laksberg, E. (1978) 'On the sensitivity analysis of linear active networks.' *Electronics Letters,* **14**(7), 221–223

Larson, H. J. (1969) *Introduction to probability and statistical inference.* John Wiley, Chichester

Leung, K. H. and Spence, R. (1974) 'Efficient statistical circuit analysis.' *Electronics Letters,* **10**, 360–362

Leung, K. H. and Spence, R. (1975) 'Multiparameter large change sensitivity analysis and systematic exploration.' *IEEE Trans. Ccts. Sys.,* **CAS-22**(10), 796–804

Leung, K. H. and Spence, R. (1977) 'Idealized statistical models for low cost linear circuit yield analysis.' *IEEE Trans. Ccts. Sys.,* **CAS-24**(2), 62–66

Lightner, M. R. (1979) *Multiple criterion optimization and statistical design,* PhD thesis, Carnegie-Mellon University

Maratos, N. G. (1984) *A new method for design centering that makes use of sensitivities,* Internal Report, Department of Electrical Engineering, Imperial College, London

Maratos, N. G. (1988) 'Algorithm for design centering based on use of sensitivity information.' *Proc. IEE, Pt.G,* **135**, 11–18

Nassif, S. R., Strojwas, A. J. and Director, S. W. (1986) 'A method for worst-case analysis of integrated circuits.' *IEEE Trans. Computer-Aided Design,* **CAD-5**(1), 104–113

Nelder, J. A. and Mead, R. (1965) 'A simplex method for function minimization.' *Computer J.,* **8**, 308–313

Ogrodski, J., Opalski, L. and Styblinski, M. (1980) 'Acceptability regions for a class of linear networks.' *Proc. IEEE Int. Sym. Ccts. Sys.,* Houston 1980, pp. 187–190

Pinel, J. F. and Roberts, K. H. (1972) 'Tolerance assignment in linear networks using non-linear programming.' *IEEE Trans. Ccts. Theory,* **Ct-19**(5), 475–479

Pinel, J. F. and Singhal, K. (1977) 'Efficient Monte Carlo computation of circuit yield using importance sampling.' *IEEE Proc. Int. Symp. Ccts. Sys.,* 1977, pp. 575–578

Rankin, P. J. (1982) 'Statistical modelling for integrated circuits.' *Proc. IEE, Pt.G*, **129**, 186–191

Rankin, P. J. and Soin, R. S. (1981) 'Efficient Monte Carlo yield prediction using control variates.' *Proc. IEEE ISCAS*, Chicago, May 1981, pp. 143–148

Roe, P. H. and Seth, K. H. (1971) 'Approximation of network response statistical moments in terms of component moments.' *Proc. 14th. Midwest Symp. Cct. Theory*, Denver, Colorado, May 1971

Rubinstein, R. Y. (1981) *Simulation and the Monte Carlo method.* John Wiley, Chichester

Scott, T. R. and Walker, T. P. (1976) 'Regionalization: a method for generating joint density estimates.' *IEEE Trans. Ccts. Sys.*, **CAS-23**(4), 229–234

Seth, A. K. and Roe, P. H. (1971) 'Selecting component tolerances for optimum circuit reproducibility.' *IEEE Proc. Int. Symp. Ccts. Sys.*, London, 1971, pp. 105–106

Singhal, K. and Pinel, J. F. (1981) 'Statistical design centering and tolerancing using parametric sampling.' *IEEE Trans. Ccts. Sys.*, **CAS-28**(7), 692–702

Soin, R. S. and Rankin, P. J. (1985) 'Efficient tolerance analysis using control variates.' *IEE Proc., Pt.G*, **132**(4), 131–142

Soin, R. S. and Spence, R. (1980) 'Statistical exploration approach to design centering.' *Proc. IEE, Pt.G*, **127**(6), 260–269

Spence, R., Gefferth, L., Ilumoka, A. I., Maratos, N. and Soin, R. S. (1980) 'The statistical exploration approach to tolerance design.' *Proc. IEEE Int. Conf. Ccts. Computers*, Oct. 1980, New York, pp. 582–585

Spoto, J. P., Coston, W. T. and Hernandez, C. P. (1986) 'Statistical integrated circuit design and characterization.' *IEEE Trans. Computer-Aided Design*, **CAD-5**(1), 90–103

Stein, M. L. (1986) 'An efficient method of sampling for statistical circuit design.' *IEEE Trans. Computer-Aided Design*, **CAD-5**(1), 23–29

Styblinski, M. A. (1986) 'Problems of yield gradient estimation for truncated probability density functions.' *IEEE Trans. Computer-Aided Design*, **CAD-5**(1), 30–38

Tahim, K. S. and Spence, R. (1979) 'A radial exploration approach to manufacturing yield estimation and design centering.' *IEEE Trans. Ccts. Sys.*, **CAS-26**(9), 768–774

Van der Waerden, B. L. (1969) *Mathematical Statistics.* George Allen and Unwin

Wehrhahn, E. (1984) 'A cut algorithm for design centering.' *Proc. IEEE Int. Sym. Ccts. Sys.*, Montreal, 1984, pp. 970–973

Wehrhahn, E. (1987) 'New simple heuristic algorithms for yield maximisation.' *Proc. IEEE Int. Symp. Ccts. Sys.*, 1987, pp. 816–819

Wehrhahn, E. and Spence, R. (1984) 'The performance of some design centering methods.' *Proc. IEEE Int. Sym. Ccts. Sys.*, Montreal 1984, pp. 1424–1438

Yang, P., Hocevar, D. E., Cox, P. F., Machala, C. and Chatterjee, P. K. (1986) 'An integrated and efficient approach for MOS VLSI statistical circuit design.' *IEEE Trans. Computer-Aided Design*, **CAD-5**(1), 5–14

Index